102nd International Scientific and Technical Conference “Intelligent Systems of Driver Assistance: Development, Research, Certification”

IOP Conference Series: Materials Science and Engineering
Volume 386

Nizhni Novgorod, Russia
18 – 19 April 2018

ISBN: 978-1-5108-6820-5
ISSN: 1757-8981

Printed from e-media with permission by:

Curran Associates, Inc.
57 Morehouse Lane
Red Hook, NY 12571

Some format issues inherent in the e-media version may also appear in this print version.

Printed by Curran Associates, Inc. (2018)

For permission requests, please contact the Institute of Physics at the address below.

Institute of Physics
Dirac House, Temple Back
Bristol BS1 6BE UK

Phone: 44 1 17 929 7481
Fax: 44 1 17 920 0979

techtracking@iop.org

Additional copies of this publication are available from:

Curran Associates, Inc.
57 Morehouse Lane
Red Hook, NY 12571 USA
Phone: 845-758-0400
Fax: 845-758-2633
Email: curran@proceedings.com
Web: www.proceedings.com

Table of contents

Volume 386

The 102nd International Scientific and Technical Conference"Intelligent Systems of Driver Assistance: Development, Research, Certification"

18–19 April 2018, Nizhni Novgorod, Russian Federation

Accepted papers received: 4 July 2018
Published online: 30 July 2018

Preface

Papers

IOP Conf. Series: Materials Science and Engineering **386** (2018) 011001 doi:10.1088/1757-899X/386/1/011001

The Automotive Engineers Association (AAE) is an independent public association which was formed in September 1991. The necessity of such institution was determined by the change in the management structure of the economy of the country in general and the automotive industry in particular. The branch institutes, that generally carried out scientific and technical studies, were weakened or lost. It led to a great need of an organization which provided an opportunity for specialists of various levels and areas to communicate with each other and engage in work together.

In October 1990, a constituent conference was held with representatives of 30 organizations, including automobile factories, research institutes, higher education institutions, etc. The Association's Board and the President were elected in that day. On September 25, 1991, the AAE was registered with the Ministry of Justice of the USSR as an all-Union public organization. In 1992, the AAE was admitted to the International Federation of Automotive Engineering Societies - FISITA. Nowadays, the AAE is a member of the Chamber of Commerce and Industry of the Russian Federation (CCI RF) and the European Automotive Engineers Community (EAEC). On September 30, 1993, the Association was re-registered with the Ministry of Justice of the Russian Federation.

The 102nd International Technical Scientific Conference of Automotive Engineers Association (AAE) "Intelligent drivers assistance systems: development, research, certification" took place in Nizhny Novgorod State Technical University (NNSTU), Russia, April 18-19, 2018. The main objective of the conference was knowledge exchange, discussion of problems and inspiring cooperation in development, research, production and operation of modern vehicles with intelligent drivers assistant system. Russian organizations were among the participants of the conference together with representatives from Belarus, Armenia, Ukraine and Uzbekistan.

IOP Conf. Series: Materials Science and Engineering **386** (2018) 011001 doi:10.1088/1757-899X/386/1/011001

More than 180 people from Russia and abroad visited the conference during those two days. Among the organizations who were involved in it were (список компаний). Also many scientists and PhD students from leading Russian universities were actively engaged in the event. The universities mentioned were (список универов).

Topics and questions discussed during the conference appeared pressing and needed further research and discussion not only as a part of AAE events but also other conferences, symposiums and forums related to research and development of ADAS and autonomous driving vehicles.

Peer review statement

All papers published in this volume of *IOP Conference Series: Materials Science and Engineering* have been peer reviewed through processes administered by the proceedings Editors. Reviews were conducted by expert referees to the professional and scientific standards expected of a proceedings journal published by IOP Publishing.

IOP Conf. Series: Materials Science and Engineering **386** (2018) 012001 doi:10.1088/1757-899X/386/1/012001

A dynamic model of unsupported pit traversal by a vehicle with 6x6 wheel arrangement

A V Papunin, V V Belyakov, V S Makarov, A A Anikin, U Sh Vahidov

Nizhny Novgorod State Technical University n.a. R.E. Alekseev, Minin st., 24, 603950, Nizhny Novgorod, Russia

E-mail: makvl2010@gmail.com

Abstract. The paper deals with the problem of unsupported pit traversal by a vehicle with 6x6 wheel arrangement. The paper contains passages from an unsupported pit traversal test report of "Korsak" wheel chassis. Typical test episodes are described. A design schematic is provided. Assumptions are provided. For the first time, a mathematical model is provided describing a dynamical pit traversal under recognition of the soil parameters, mass and inertia chassis specifications, as well as the travel speed. A conclusion is made that the obtained model is a more generalized case of existing models describing profile obstacle negotiation by wheel vehicles.

1. Introduction

While crossing complicated terrains it is important for a vehicle to retain sufficient mobility to succede in its functional operations. The mobility is an integral operational feature of a Special Purpose Vehicle (SPV) determining its performance specifications with optimum adaptivity to the operation conditions and the technical state of order of the vehicle itself, that is the vehicle's capability of coping with external and internal factors impeding the mission progress [1, 2].

While crossing in spring and in summer terrains susceptible to flooding, also in floodplains after the water level has gone back, vast complicated landscapes may be observed formed by numerous brooks and representing a challenge for an SPV. As a rule, the resulting profiled obstacle may from the point of view of SPV mobility classified as pits, scarps and counterscarps. Thereby, all these obstacles are unsupported and crumbling. The most complicated case is a traversal of unsupported pit [3]. The mobility of an SPV at unsupported pit traversals was the subject of earlier papers of the authors hereof [3-5]. That papers describe pit wall crumbling patterns depending upon the parameters od the soil and the chassis specifications. Also, the dimensions of the pit to overcome correlate with the vehicle's dynamics. Therefore, we further consider a dynamical model of a pit traversal by a vehicle of 6x6 wheel arrangement.

2. Assumptions

We consider a model of pit traversal by a multi-axle vehicle. The equation of the chassis motion obtains the following appearance:

$$\sum_i F_{\varphi i} - \sum_j F_{fj} = m\frac{dV}{dt},$$

Thereby, $\sum_i F_{\varphi i}$ is the sum of traction on the chassis wheels, $\sum_j F_{fj}$ is the sum of resistances, m is the chassis weight, $\frac{dV}{dt}$ is the acceleration.

Simultaneously, during the pit traversal, the chassis shall follow a rotation trajectory while nosediving into the pit. The equation will then obtain the following appearance:

$$\sum M = J\frac{d\omega}{dt},$$

Thereby, $\sum M$ is the sum of the moments of forces acting on the chassis, J is the moment of inertia, $\frac{d\omega}{dt}$ is the angular acceleration.

While driving with a variable speed, the diving motion of the front axle of the chassis is determined by the dimensions of both the vehicle and the pit, the values of the soil, as well as by the driving speed. Important thereby are the axle load distribution and the location of the centre of gravity, the wheel arrangement over the wheel base and the number of wheels. Depending upon these parameters, the pit traversal diagram of a multi-axle vehicle will vary.

Thereby we introduce a number of assumptions, a part of which is aligned with the design parameters of the test chassis unit.

1. The axle load distribution between the left and the right wheel of one axle is even.
2. When solving the pit traversal problem, we assume that the resultant of the forces occurring due to the wheel-terrain interaction is concentrated in the centre line of the wheel passing via its centre of rotation.
3. The location of the centre of gravity does not vary during the vehicle motion.
4. The positions of the rotation axes of the wheels relative to the chassis centre of gravity remains stable.
5. The influence of the tire stress which changes at dynamic loads occurring during the pit traversal and the re-distribution of the reactions over the drivers are neglected or assumed constant.
6. We also assume a constant pit tranversal speed.

3. Pit traversal diagram

We consider based on an example of a 6x6 chassis a pit traversal diagram and the typical components of the motion. The figure 1 (left) demonstrates the episodes of a test carried out October 21, 2018 with "Korsak" SPV with 6x6 wheel arrangement. The relevant diagrams are shown in the right column.

Fig. 1 shows typical pit traversal stages. Considered in detail, Fig. 1 a shows the vehicle starting the pit traversal crushing the front pit wall. The load is distributed over all the wheels of the vehicles. The diagram is shown in Fig. 1 b.

Fig. 1 c shows subsequent vehicle motion. A forward nosedive of the vehicle takes place. Thereby, the nosedive amplitude will depend upon the weight, the dimensions, the inertia of the vehicle, on the speed and on the pit dimensions. The chassis motion diagram is shown in Fig. 1 d. Thereby, the locomotion of the vehicle can be described by equations of forward and rotary motion.

Fig. 1 e shows vehicle's thrusting against the crushed pit edge, thereby a significant vehicle tilt is obvious. Fig. 1 f provides a related diagram. The pit traversing capacity of the vehicle will be determined by the diagram shown in the figure. The locomotion condition will be a case when the sum of torques CW is higher than that CCW. A critical motion condition is the equality of the torques. Thereby, the critical pit width can be determined.

Fig. 1 g shows the moment the chassis has climbed the opposite pit wall. Fig. 1 h shows the diagram. Further, Fig. 1 i shows the pit left behind by the wheels of the 1st and the 2nd axle. Fig. 1 j shows the relevant diagram. In principle, a sort of "taildive" may occur thereby. But for a chassis with forward-shifted centre of gravity such phenomenon is deeming negligible at a pit traversal, the condition whereof is assumed in accordance with Fig 1 f. This thesis is based on the fact that as a rule multi-axis chassis are similar to the one considered, with the centre of gravity displaced to the front. It should be stressed, that the approach to the calculation of the maximum width of traversed pit for vehicles with the centre of gravity displaced aft from the middle axle is similar to that set forth above.

It should be noted, that pendular chassis swings are inevitable during the pit traversal due to fluctuations of different types. This phenomenon is not accounted for in the presented model.

IOP Conf. Series: Materials Science and Engineering **386** (2018) 012001 doi:10.1088/1757-899X/386/1/012001

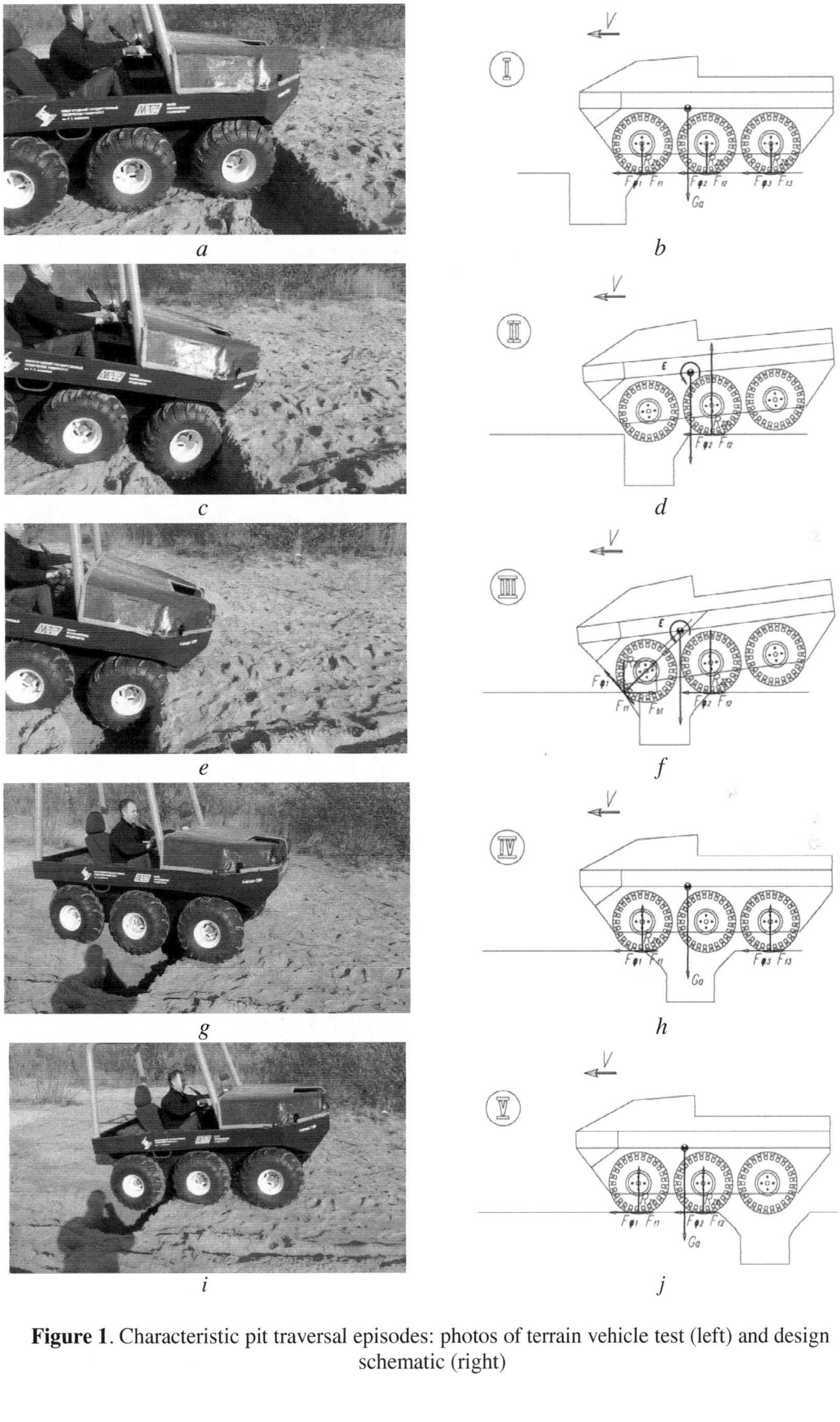

Figure 1. Characteristic pit traversal episodes: photos of terrain vehicle test (left) and design schematic (right)

4. Mathematical model

In accordance with the assumptions made, we consider in a detailed manner the traversal diagram of a pit with critical width (Fig. 2).

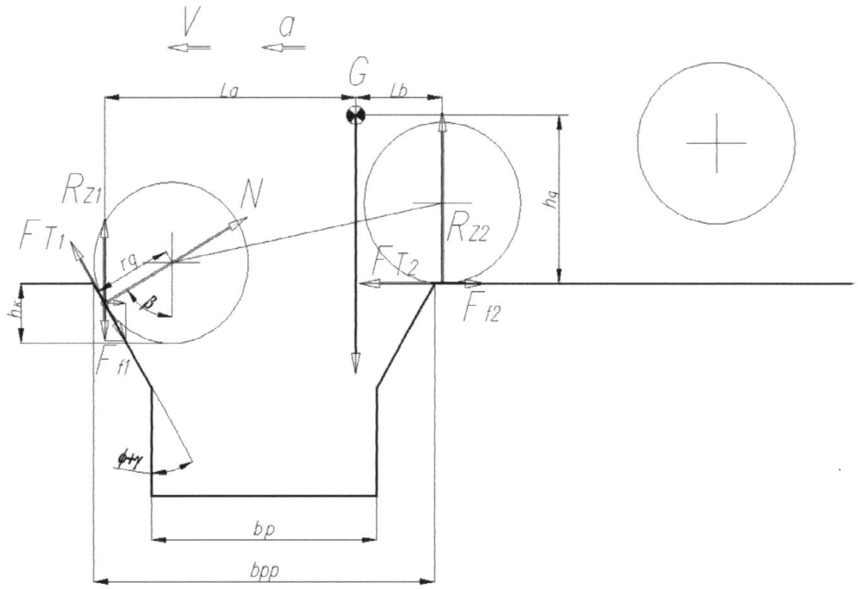

Figure 2. Calculation diagram of the critical width of unsupported pit

We now consider thoroughly the diagram in Fig. 2.

The nosedive calculation is carried out in accordance with the rotary motion equation. In a general case, the latter will look like:

$$J\varepsilon = G_a l_b - F_w h_w - F_{\text{кр}} h_{\text{кр}} - m_a a h_g, \tag{1}$$

Thereby, J is the moment of inertia of the chassis, ε is the rotary acceleration, G_a is the machine weight, l_b is the moment arm, $l_b = l_2 \cos\alpha$, l_2 is the distance of the centre of gravity from the axis of the 2nd axle, α is the nosedive angle, F_w is the air resistance, h_w is the height of the centre of sail, $F_{\text{кр}}$ is the resistance of the hook load, $h_{\text{кр}}$ is the hook load application height, m_a is the vehicle's mass, a is the acceleration of the chassis, h_g is the height of the centre of gravity.

In case of a steady low-speed locomotion at small nosedive amplitude, we can assume that:

$$\varepsilon = \frac{G_a l_2}{J}. \tag{2}$$

The nosedive amplitude may vary depending on the pit width: it will be determined in accordance with the following motion equation.

$$\alpha = \alpha_0 + \omega_0 t + \frac{\varepsilon t^2}{2}, \tag{3}$$

Thereby, α_0 is the initial angle, $\alpha_0 = 0$, ω_0 is the initial angular speed, $\omega_0 = 0$, t is the pit traversal time till the opposite wall contact, $t = \frac{S}{V}$, S is the pit width measured from the crushed front to the contact with the wall.

Thus, the nosedive amplitude can be calculated in accordance with the following equation:

$$h_{\text{к}} = (l_1 + l_2)\sin\alpha, \tag{4}$$

Thereby, l_1 is the distance of the centre of gravity to the 1st chassis axle.

We consider the correlation of parameters in accordance with the diagram in Fig. 2.

$$G_a = R_{z1} + R_{z2}, \tag{5}$$

$$\frac{R_{z1}}{R_{z2}} = \frac{l_a}{l_b}, \tag{6}$$

$$l_a = l_1 \cos\alpha + r_{\text{д}} \sin\beta, \tag{7}$$

Thereby, $r_{\text{д}}$ is the dynamic wheel radius.

The angle β will be determined based on the following dimension:

$$\beta = \min\left[90 - (\varphi + \gamma); \arccos\left(1 - r_д h_к^{-1}\right)\right], \tag{8}$$

Thereby, φ is the repose angle of the soil, γ is the inclination angle of the pit wall [6].

For evaluation of the motion options, we consider the chassis motion equation. We project the forces on the OX, OY axes and consider the equality of the moments in relation to the contact point of the front wheel with the pit wall.

$$\begin{gathered} X: F_{T_1} cos\beta - F_{f_1} cos\beta - N sin\beta + F_{T_2} - F_{f_2} - F_w - F_{кр} = m\frac{dV}{dt} \\ Y: F_{T_1} sin\beta - F_{f_1} sin\beta - N cos\beta + R_{z_2} - G_a = 0 \\ M: -G_a l_a + R_{z_2}(l_a + l_b) + \left(R_{x_2} - F_{f_2}\right)\left(h_k - \left(r_д - r_д cos\beta\right)\right) + \\ + ma\left(h_g - \left(r_д - r_д cos\beta\right)\right) - F_w\left(h_w - \left(r_д - r_д cos\beta\right)\right) = J\frac{d\omega}{dt} \end{gathered} \tag{9}$$

Thereby, F_{Ti} is the employed traction force on the wheels of the ith axis, F_{fi} is the resistance on the wheels of the ith axis, N is the normal response of the pit wall to the wheel.

The maximum width of a traversed pit will correspond to a traction force on the wheels not in exceed of the cohesion force. Thus, the traction force can be calculated as follows:

$$F_{T_i} = \left(c_ш A + N_i \mathrm{tg}\varphi_ш\right)k_н + (cA + N_i \mathrm{tg}\varphi)(1 - k_н), \tag{10}$$

Thereby, $k_н$ is the coefficient of pattern area saturation, $c_ш$, $\varphi_ш$ are the cohesion modulus and the friction angle of the tire tread with the soil, c is the inherent soil cohesion.

Thus,using the formulae (1) – (10) the maximum traversed width of the unsupported pit by a vehicle with 6x6 wheel arrangement.

5. Conclusion and future work

A study of profile passability was carried out, specifically, a traversal of an unsupported pit by a vehicle with 6x6 wheel arrangement. Terrain experiments were conducted with a “Korsak” SPV with 600 kgs weight.

Typical test episodes were illustrated. A diagram was developed for mathematical modelling of a pit traversal by a wheel-type vehicle.

A mathematical model was developed for calculation of an unsupported pit width dynamically traversed by a multi-axle wheel-type vehicle.

The analysis of the obtained dependencies has demonstrated the following:

- the dependencies obtained are valid for a dynamic traversal of an unsupported pit.
- if the pit is of hard rock type then the obtained dependencies are transformed to the problem of hard pit traversal.
- if the problem is considered statically then the formulae are transformed to dependencies for climbing of a hard step.

Thus, the obtained dependencies are not contradicting the dependencies obtained earlier by other authors [3-5] being a more general case describing the overcoming of threshold type obstacles.

References

[1] Belyakov V V, Belyaev A M, Bushueva M E, Vahidov U Sh, Goncharov K O, Zezyulin D V, Kolotilin V E, Leliovsky K Y, Makarov V S, Papunin A V, Tumasov A V and Fedorenko A V 2013 The concept of mobility of surface transport-technological vehicles *Proceedings of the NGTU. R.E. Alekseeva* **3 (100)** 145–174

[2] Makarov, V., Zeziulin, D., Belyakov, V. Prediction of all-terrain vehicles mobility in snowscape scenes. 18th International Conference of the International Society for Terrain-Vehicle Systems, ISTVS 2014, South Korea, 2014

[3] Galkin A A 2012 Mathematical model of overcoming the destroyed moat by a multiaxial wheeled vehicle *Journal of Automotive Engineers* **6** 40-42

[4] Agyikin Ya S 1981 Passability of cars *Mechanical Engineering* 230

[5] Agyikin Ya S 1972 Off-road wheeled and combined propellers. Theory and calculation *Mechanical engineering* 184

[6] Tsytovich N 1963 Soil mechanics. Edition 4 *Stroyizdat* 638

Analysis of the main bases of the coastal zone, as a pathway for the movement of special vehicle and monitoring complexes

A M Belyaev[1], A A Zakharov[2], V S Makarov[1], V A Zuyev[1], V N Kravets[1]

[1]Nizhny Novgorod State Technical University n.a. R.E. Alekseev, Minin st., 24, 603950, Nizhny Novgorod, Russia
[2]CJSC Concern Termal, Gagarin Avenue, 178, 603950, Nizhny Novgorod, Russia

E-mail: makvl2010@gmail.com

Abstract. The key point of the article, developed in Nizhny Novgorod State Technical University n.a. R.E. Alekseev is Autonomous mobile robotic complex for monitoring of coastal zones. The classification of coastal zones as the pathway for special vehicle and mobile complexes of coastal zones monitoring is considered. The analysis of traffic conditions is carried out. It is concluded that the dominant in the coastal zones, which account for most of the supporting surfaces, can be considered sand-gravel and sand areas, as well as snow pathway in the winter.

1. Introduction

About 40% of The world's population lives on a narrow coastal strip along the oceans and seas. Coastal zones of the Russian Federation occupy about 2/3 of its borders and about 10% of the coastal zones of the land surface of the Earth. In addition, most of the population lives along the coastal zones of inland water bodies, in particular, lakes, ponds, rivers, streams, artificial reservoirs. For many observations, the coastal zone is the boundary area of land and water environment, consisting of the shore itself, determined by the level of water fluctuations at tides and tides, the coastline and the coastal slope. The width of the coastal zone can be from several tens to several hundred meters. The main factors that determine the nature of the coastal zone-geological, climatic, engineering human activities, etc. The coastal zone is characterized by temporary instability and changes its position depending on the time of year, storm conditions, the phase of the tide, atmospheric pressure and many other reasons.

Coastal zones and coastal territories are the place of residence for large number of people and the areas of their cultural activity. As a result, the accompanying industries, forest, agriculture and fish farms are widely spread there. The production of natural minerals, oil and gas, ores also take place in such areas. The alternative power engineering is also developed there. The coastal zones were always the objects for tourism and rest and the research activity areas. Therefore, the safety of human activity, state and ecological (technogenic and natural) safety for these territories is an important and relevant task.

The monitoring of the coastal territories and adjacent water areas in a wide spatio-temporal framework is necessary for the safety control. Monitoring can be implemented by means of a large number of stationary posts, or by the means of the mobile complexes. The mobile land complexes for monitoring of coastal zones look like the most reasonable solution now. An Autonomous mobile robotic complex (AMRC) for monitoring coastal zones was developed as part of the project conducted at NNSTU n.a. R. E. Alekseev [1, 2]. Chassis of AMRC is shown in Fig. 1.

Fig 1. Fragment of the AMRC test.

Special operating conditions, which are not typical for conventional land-based complexes, determine the development of methods and research to obtain characteristics of road and ground surfaces in the coastal zone. Therefore, the analysis was carried out, the classification was developed and the dominant surfaces in the coastal zones were identified.

2. Classification of the coastal zones

The analysis made it possible to develop a classification of coastal zones as a pathway for transport technological vehicles (TTV) and mobile complexes of coastal zones monitoring.

From the perspective of Geology, [3, 4] the following structures are distinguished: morphological-steep banks, with a bar, low — lying, etc.; structural-longitudinal, transverse, diagonal banks; dynamic, taking into account the initial relief, bottom slopes, wave intensity and direction, geological structure and sediment balance; genetic, which is based on various principles [3, 4].

From the point of view of the movement of TTV support surfaces are classified according to: microprofile determining the hilly, microprofile important for assessing the smoothness of the course and the bearing and shear characteristics of the soil determining the possibility of movement of any vehicle. Coastal zones are distinguished by the size of the water body - a river, lake, sea ocean.

The type of soil is the next classification feature. The most typical are like sand (different grain size), clay (differ in plasticity), rocky (rocky, vary according to the size of a pebble), mixed.

From the point of view of the presence of plants areas with vegetation and without can be identified. Moreover, flora can be different, so on the shore areas can grow shrubs and fast-growing trees, representing some difficulty for the movement of TTV, and in some areas there are only herbaceous plants that do not present difficulties for movement.

In winter, Snow, ice, sludge, ice are the difficulty to move in winter.

The entrance to the coastal area is also important. You can select areas with the possibility of descent, hard to reach and impossible to descend for the vehicle. For the latter, the possibility of movement is determined by the presence of adjacent sections in the border areas.

The movement of TTV and AMRC is almost always carried out with some roll. These modes will be determined by the longitudinal angle of the coastal zone. Therefore, you can select areas with a large transverse slope and traffic on the road without a transverse slope.

You can highlight areas on the uniformity of the characteristics of the movement from the point of view of the movement of AMRC. This is the constancy (within certain limits) of the traction characteristics of the surfaces of motion, the characteristics of macro and microfile, including the presence of flooded areas and bays.

As the humidity of the soil surface approaches the water, this classification should also vary accordingly.

The characteristics considered will be variable in time with some periodicity, for example, day or season. Variability during the day is determined by tidal activity.

From the point of view of the formation of characteristics and mainly the geometry of the shore, natural and anthropogenic (man-made) factors can be identified.

The structure and classification features are shown in the Figure 2.

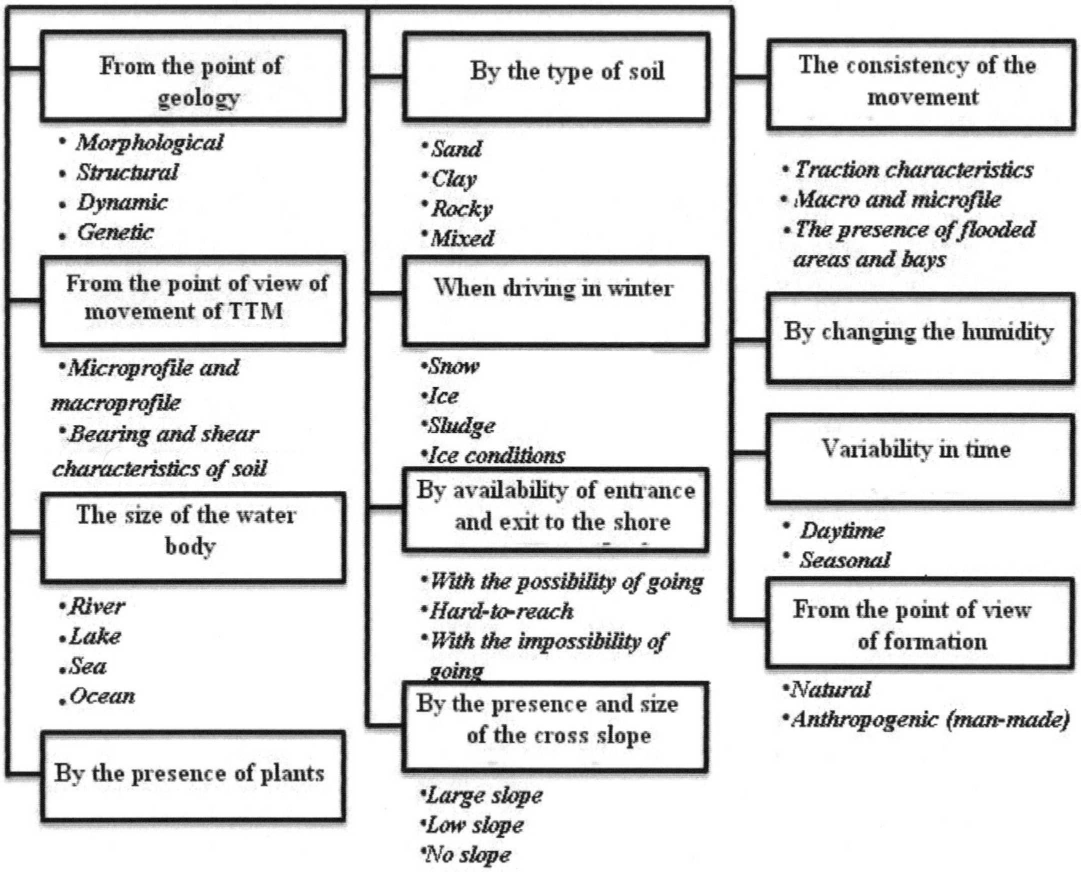

Fig 2. Structure and classification of coastal zones

The use of this classification will allow to systematize the data on coastal zones as the pathway for TTV and AMRC, which in turn will give the opportunity to predict their mobility [5].

Despite the variety of supporting surfaces of coastal zones presented in this classification, to determine the basic parameters of the chassis is enough to know only the dominant conditions of movement.

3. Analysis of dominant surfaces in coastal zones

The coastal zone is composed of Sands, pebbles (accumulative phases), coarse clastic and crushed material (abrasive phase) and sediments brought by rivers and winds from the depths of continents to the sea coast.

The abrasive shore is a high steep receding shore of the reservoir, destroyed by the action of the surf. The main elements of the relief of this type of shore are: abrasive underwater slope (bench), coastal ledge (cliff), wave-hole and underwater abutment terrace. Accumulative shore is coming the shore, built with sediments brought by waves and surf..

Potential traffic surfaces of TTV and MARK as the dominant traffic area are beaches or coastal terrace. Movement on coastal terraces in most cases is no different from the movement of cars on conventional soil surfaces. The beaches of coastal zones are of the greatest interest as the surface of the movement due to the fact that their physical and mechanical properties can change with high frequency during the day, due to the instability of climatic conditions and the action of tides in the coastal zone.

The coastal zone has unique reserves of various mineral materials, primarily sand and gravel [6]. The analysis of materials composing the support base for the motion of TTV and AMRC chassis in the coastal zone can be limited to the consideration of sand and gravel materials.

Since the use of transport and technological tasks in the coastal zone is carried out year-round, in regions with a cold climate, sand-gravel and stone bases in the winter period are covered with ice and snow.

The analysis of the conditions of movement of TTV and AMRC conducted in the works of the authors of the article, as well as data from the works [7-9] showed that it is possible to cut out the dominant surfaces in these areas, namely sand and gravel areas, as well as snow pathway in the winter

4. Conclusion

According to the results of the study described in the article, the following conclusions can be made:

- The topicality of development of special ground mobile complexes of coastal zones monitoring is shown. An example of an Autonomous mobile robotic complex (AMRC) developed with the direct participation of the authors is given.
- The classification of routes in coastal zones as the pathway for transport and technological machines is developed. The classification features of coastal zones associated with Geology, theory of vehicle traffic, with the size of the water body, type of soil, the presence of plants, taking into account the movement in the winter, the presence of the entrance and exit to the shore, the presence and size of the cross-slope of the shore, with the constancy of the characteristics of movement, humidity change, taking into account the variability of characteristics in time, taking into account the creation.
- It is established that in coastal zones dominant supporting surfaces are sand-gravel and sand areas, as well as snow pathway in the winter season.

References

[1] Belyakov V.V, Beresnev P. O., Zeziulin D. V., et al. 2016 Autonomous Mobile Robotic System for Coastal Monitoring and Forecasting Marine Natural Disasters Proceedings of the Scientific-Practical Conference "Research and Development - 2016" DOI 10.1007/978-3-319-62870-7_14

[2] Kurkin A., Makarov V., Zeziulin D., et al. 2017 Study of coastal soil surfaces of Sakhalin Island *Proceedings of the Thirteenth International MEDCOAST Congress on Coastal and Marine Sciences, Engineering, Management and Conservation, MEDCOAST 2017* 795-804

[3] Kozlovsky E.A., Ledovskikh A.A. 2010 Russian geological encyclopedia *VSEGEI* 663

[4] Classification of coastal *http://dic.academic.ru/dic.nsf/enc_geolog/*

[5] Belyakov V V, Belyaev A M, Bushueva M E, et al. 2013 The concept of mobility of surface transport-technological vehicles *Proceedings of the NGTU. R.E. Alekseeva* **3 (100)** 145–174

[6] Makarov V., Filatov V., Vahidov U., Kurkin A. and Belyakov V. 2017 Study of Trafficability Conditions of Typical Soils of Coastal Zones of Sakhalin Island (Russian Federation) *Proceedings of the 19th International & 14th European-African Regional Conference of the ISTVS*

[7] Stepanov A. P. 1983 Designing and calculation of floating cars *Mechanical engineering* 198

[8] Redkin M. G. 1966 Floating wheel and tracked vehicles *Voyenizdat* 200

[9] Malakhov D. Y. 2009 Development of a technique of assessment of hydrodynamic impact on the floating cars entering a surf zone *Thesis Cand.Tech.Sci.* 202

IOP Conf. Series: Materials Science and Engineering **386** (2018) 012003 doi:10.1088/1757-899X/386/1/012003

Autonomous electric commercial vehicle for difficult operating conditions

D V Zeziulin[1], D Yu Tyugin[1], K M Shashkina[1], V A Kuzmichev[2], P O Beresnev[1]

[1] Nizhny Novgorod State Technical University n.a. R.E. Alekseev, Minin str., 24, Nizhniy Novgorod, 603950, Russian Federation
[2] Peter the Great St. Petersburg Polytechnic University

E-mail: pavel.beresnev@nntu.ru

Abstract. Nowadays development of autonomous vehicle driving control systems is a global trend. The growing autonomous vehicles market is stimulated by a variety of competitions and awards. This article introduces the background and the first steps taken by the research team of Nizhny Novgorod State Technical University named after R.E. Alekseev (NGTU). The research is supported by GAZ Group's United Engineering Centre and by PAO GAZ experts. The control system in question is developed for operation in severe weather and road conditions of the Russian Federation. An electrical platform is used as chassis for the control system. The working components of the autonomous vehicle driving control system, its location and the area of operation are considered in the article. The software to operate every module of the system has been developed; the system operation has been tested.

1. Introduction

Autonomous vehicle driving control systems market is a relatively new and yet rapidly growing field both in Russia and globally. A number of large motor manufacturers including IT-companies, such as Google [6], Nvidia [7], Baidu [8], and military machinery manufactures DARPA [9], Oshkosh Defense [10], Lockheed Martin [11], etc. (Figure 1) are developing their autonomous vehicle driving control systems based on advanced driver assistance systems (ADAS). as well as some scientists [1,2,3,4,5], IT-companies, such as Google [6].

a b c d

Figure 1. (a) Baidu self-driving car[8] , (b) Oshkosh unmanned ground vehicle [10], (c) Google self-driving car [6], (d) Nvidia self-driving car [7]

DARPA Grand Challenge [9] autonomous vehicle contests sponsored by the US Government helped evaluate the performance and enhance competition in the autonomous vehicle driving control systems market. The contests revealed all the shortcomings of autonomous vehicles and influenced considerably the autonomous vehicles development. It should be pointed out that most of the contestants were represented by joint efforts of a large motor manufacturer, which provides a vehicle, and a university research team, developing an autonomous driving system for the vehicle.

In 2014 the US set the first national standard for autonomous vehicles SAE J3016 «Taxonomy and Definitions for Terms Related to On-Road Motor Vehicle Automated Driving Systems» [7]. This standard defines autonomous vehicles and provides the taxonomy for motor vehicle automation. The report lays the groundwork for further development of the standards and demonstrates high level of cooperation between the autonomous and automated vehicles developers.

The report states that Google has reached the highest level of motor vehicle automation Level 5 being Full automation.

The main difference between Google's and other companies' products is that Google immediately verifies the traffic data obtained by LiDAR while driving with 3D high resolution maps included in the system earlier. This allows comparing the data to detect any possible obstacles on the route. That is why for now it is only possible to follow the routes which have detailed 3D maps.

The Russian Federation cannot boast a highly successful development of autonomous vehicle driving control systems, however such companies as Yandex [13], Cognitive technologies [14], FGUP NAMI [15], PAO KamAZ [16], Avrora robotics [17] and others have given the field a good start.

Yandex [13] is one of the most successful autonomous vehicle driving control systems developers in Russia. The researchers aim to introduce self-driving cars into transportation services (taxi, haulage, etc.). Yandex has by now designed two self-driving car prototypes using Toyota Prius and Kia Soul as the base. The hardware includes the following sensors. Inside the car is equipped with front- and rearview cameras, which provide data to detect and identify any vehicles, people, traffic signs and road surface marking as well as the driveway borders. Velodyne HDL-32 and Velodyne VLP-16 are mounted on the roof of the vehicles. These sensors use a laser emitter to scan the environment. The obtained data then becomes the basis for a 3D map, which helps calculate the exact distance to the objects around the vehicle. The car is also equipped with sensors to identify its current position, speed and trajectory, like GPS/GLONASS, an inertial measurement unit and odometric sensors.

To reduce the gap in the technological development of autonomous vehicle driving control systems designed specifically for Russian weather and road conditions NSTU research team and GAZ Group United Engineering Center are developing and installing autonomous vehicle driving control systems on GAZ commercial vehicles.

2. Autonomous Vehicle Driving Control System Development

We use an electrical platform for commercial vehicles as chassis to mount the control system on (Figure 2).

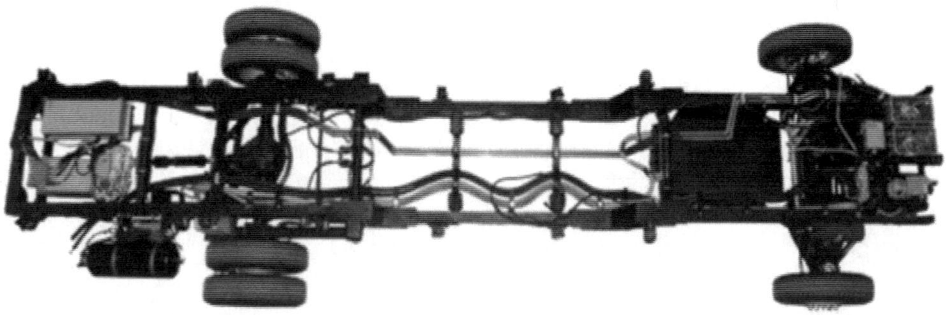

Figure 2. An overall view of the electrical platform

IOP Conf. Series: Materials Science and Engineering **386** (2018) 012003 doi:10.1088/1757-899X/386/1/012003

A number of advantages of electric vehicles over traditional ones have influenced our choice, the important one being a higher level of safety and environment friendliness of an electrical vehicle due to the use of electricity as the main source of power.

Most of the largest motor manufacturers are carrying out research in the field, while the most advanced countries like the USA, China, Japan, Germany, and the UK are engaged in long-term programs to support electric vehicle manufacturers and encourage the consumers. Right now the Russian Federation is about to introduce a program to boost the field as well.

Electrical units unlike mechanic, hydraulic or pneumatic units of the chassis do not need actuators or any drive gears to operate, which makes the vehicle systems safer and more reliable, and at the same time reduces their costs.

Although this platform is the base for a wide range of GAZ commercial vehicles, we chose a bus to implement our autonomous vehicle driving control system. The choice of a public transportation vehicle is driven by the following reasons: a preplanned path for the vehicle, following the special lane for public transport, an opportunity to recharge the batteries at bus terminals.

At the first stage of our development we defined the working components of the system and the coverage of the sensors.

The system uses an onboard computer based on INTEL Xeon Processor E5-2690 V4 CPU with three NVidia GeForce GTX 1080 Ti graphics cards to collect and process data obtained from different sensors.

LiDARs are to be used to detect any other moving objects and obstacles, as well as to create a virtual route map. Velodyne HDL-32 scans the environment in front of the vehicle. Velodyne HDL-32 is a wide vertical field of view sensor (from +10° to -30°) detecting targets at up to 200 meter range. The less important rear and lateral views are covered by Velodyne VLP-16, with vertical FOV from -15° to +15° and up to 100 meter range. Besides, the system is equipped with a planar LiDAR Sick LMS Pro 511, mounted at the bumper level of the vehicle. The choice is based on a high scanning frequency of the sensor (75 Hz) and ambient operating temperatures ranging from -30°C to +50°C. The LiDAR data is passed to the onboard computer over an Ethernet connection. We developed software for object detection: we used a modification of YOLO v2 [18] network for real-time object detection on the video stream, and point cloud clustering algorithms along with linear algebra methods for LiDAR data based object detection. All the modules are integrated in ROS (Robot operating system) [19] to ensure their smooth interaction.

In autonomous vehicle driving control system development reliability of the system as a whole is essential, especially for extreme weather and road conditions. To increase the reliability of the obstacles detection system we equip autonomous vehicles not only with LiDARs, but also with long- and short-range radars. And we suggest that Continental ARS441 and SRR510 operating at a frequency of 77 Hz are the best fit. ARS 441 operates at up to 174 meters range with a horizontal FOV of ± 10°. As for SRR510, it operates at up to 70 meter range with a horizontal FOV of ± 75°. Both of them are connected to the onboard computer through Ethernet. The system combining data from several radars allows a more detailed evaluation of the traffic situation in front of the vehicle and long-range measurements of the distance to any obstacles. Radars are less affected by weather conditions, color of the objects and light. Radars boast a high range and speed resolutions, immediate data processing ability, which is particularly important at higher speeds. It should be pointed out, that different types of radars based systems are already widely used by some of the large motor manufacturers to develop their ADAS.

It is mandatory to obey the traffic regulations while operating a self-driving car on public roads. We use a vision system and a Basler [20] camera as the main working unit to detect any road users, such as vehicles, pedestrians, traffic signs and road surface marking, etc. Our autonomous vehicle driving control system includes three cameras fully covering the forward view.

In order to increase security and efficiency of the system we use a Titan 632 [21] thermal imaging camera for obstacles detection. The camera is characterized by horizontal and vertical FOV of 90° and 65° respectively, and ambient operating temperatures ranging from -60°C to 50°C. The thermal image

IOP Conf. Series: Materials Science and Engineering **386** (2018) 012003 doi:10.1088/1757-899X/386/1/012003

processed by the onboard computer allows a better driveway borders detection, earlier evaluation of the upcoming corners and curves on the route. It is important that the thermal imaging camera unlike video cameras operates smoothly even when exposed to lighting of oncoming traffic, smoke, dust, fog or rain.

We developed a path planning software to set a path based on particular GPS coordinates and to process trajectory data into executive instructions. The software solves an optimization problem by assigning optimal executive instructions and minimizing any possible path and target speed errors.

The software involves the following modules: trajectory planning module, GPS data collecting module, data displaying module, path drawing module. The main working parts are OC-203-GSM GNSS receiver, AT330 GNSS antennas and Xsens [22] inertial unit. The path planner operates on separate integration of data obtained from GNSS receiver and IMU, meaning their performance is not dependent on each other. At regular intervals IMU is adjusted based on GNSS data. The navigation set provides highly-accurate GPS coordinates, an appropriate speed above the earth, altitude, and the quantity and quality of satellite signals. It connects to the onboard computer through Ethernet interface. Built-in software processes the data, manages the database and connects measurement data to any sensors in the system using GPS coordinates to operate.

The suggested vehicle is a kinematic model using model predictive control (MPC) schemes. A high-level logic manages the driving control during the start, while overcoming any obstacles on the route and when the target point is reached. Every step of the algorithm (meaning every time GPS receiver provides new data) solves a problem to assign optimal executive instructions to follow the path from the current location to the next preset point. Thus the system helps predict the vehicle's trajectory given the current position, speed and set of executive instructions. Optimal control parameters help specify the executive instruction set.

3. Autonomous Vehicle Driving Control System Experimental Tests

Before we mounted the vision system, which is responsible for traffic object detection, on the electric bus we tested it under real-life traffic conditions.

The tests involved detection and identification of vehicles, pedestrians, traffic signs and traffic lights. The system's software complex includes newly developed OCVStudio software based on OpenCV [23] database. At the input the complex receives an image from the optical sensor, which is then roughly processed over neural network and followed by neural network data output.

We used a YOLO v2 network modification to develop and fit the neural networks. During the development we made sure to avoid overfitting of deep neural networks, added screening layers, deepened convolutional layers and stretched fully connected layers.

The user interface of the complex displays real-time vehicles, pedestrians, traffic signs and traffic lights data. Refer to figure 3 to see some snaps from showing the testing process.

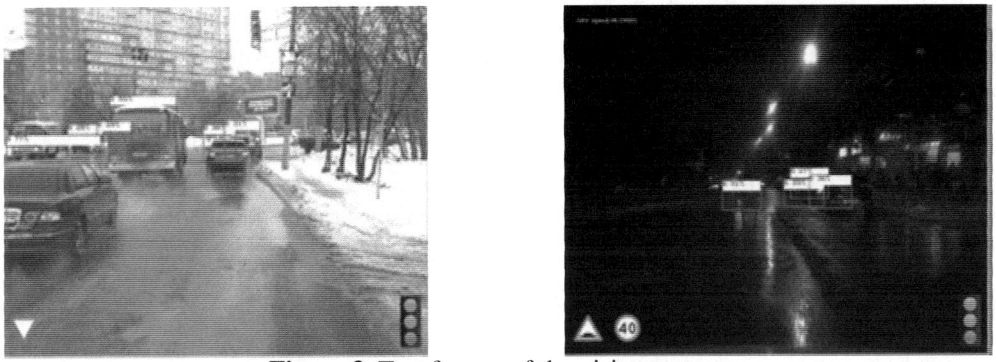

Figure 3. Test frames of the vision system

IOP Conf. Series: Materials Science and Engineering **386** (2018) 012003 doi:10.1088/1757-899X/386/1/012003

The benchmark of the proper system operation is accurate detection and identification of the traffic objects. The tests suggest that the system accuracy, showing how many of the detected traffic objects were identified correctly, goes up to 78%.

At the same time we also conducted experimental tests on detection and identification of any obstacles on the route of the vehicle as well as on localization ability and path planning, which took place on GAZ Group's specialized test site. Figure 4 shows some snaps of the tests.

Figure 4. Experimental studies of the unmanned traffic control system

LiDAR tests help evaluate the ability of the system to detect obstacles and generate a point cloud for further localization these points. Figure 5 reveals some details of the tests.

Figure 5. Fragments of tests with LIDARs

If an obstacle detected was higher than 15 cm or closer than 8 m, the vehicle was stopped. Driving was only possible with no detected obstacles in 9 m. Thus starts and stops upon an obstacle signal need a trigger to be completed, which helps avoid repeated signals if an obstacle on the border of detection zone.

The LiDAR tests reveal that obstacles are detected and identified both correctly and accurately. Besides, the tests justified the use of LiDARs to localize a vehicle. Our further step is to ensure LiDAR real-time map is compared to the preset map, which will allow localization of the autonomous vehicle in the relative coordinate system with LiDAR as a center.

The earlier mentioned GPS system data was the basis for path planning and trajectory control. A user draws a route over Tracker software interface. The vehicle follows the trajectory based on preset points and receives RTK corrections to the preplanned path from base b and executive instructions from MPC system through the GPS/GLONASS receiver. The vehicle took more than 50 rounds following the preplanned path, which simulated urban environment with buildings standing close to each other, metallic objects, farms and smokestacks. We evaluated the navigation system after every round by comparing data from GNSS receiver, IMU and preset trajectory, and by evaluating the signal quality. During the rounds we also evaluated the consistency and smoothness of the trajectory; the connection quality on the route; the connection between the executive functions control unit and the onboard computer; and the inertial unit performance. Figure 6 shows the details of the test.

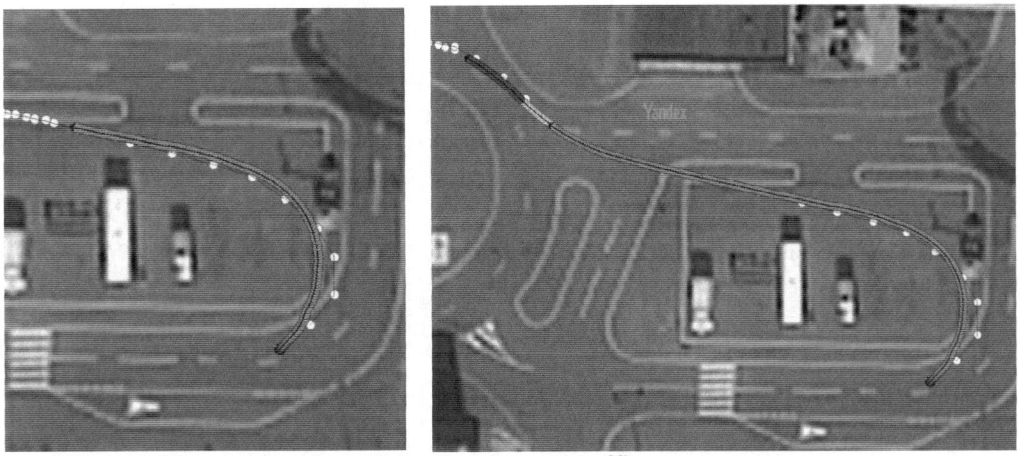

Figure 6. The user interface of the planning and traffic management program

Satellite tracking quality on the route complies with the standards, the exception being some metal trusses and smokestacks, which caused delay in trajectory update of up to 3 seconds. The signal quality was satisfactory, although at some points (especially at the start of the route) the connection could be unstable. We suggest using Wi-Fi to transfer updated data from the base to the vehicle with a backup communication channel. Optimal distance between the points along the route is about 2 meters.

It is important to note, that navigation trajectory is not an absolutely reliable source for vehicle positioning. The system performance is optimal in urban environment without tunnels or any other long-time disruptions of GNSS satellite connections. The tests reveal that INS trajectory deviation from GNSS trajectory never exceeded 3 meters, with deviation minimum being 1.334 meters, and deviation maximum being 2.647 meters. A big gap between the minimum and maximum values can be explained by a long-time GNSS connection disruption (more than 10 seconds) because of the metal trusses and smokestacks and low speed of the vehicle.

4. Conclusion

An autonomous vehicle driving control system was developed, and the working units for the system were designed. We drafted design documents for autonomous vehicle driving control system modules, for software and data communications networks, connecting all the subsystems into a single system on the onboard computer. We developed an electrical platform to mount the autonomous vehicle driving control system on. Experimental tests on different modules of the system were conducted, such as a vision system for traffic objects detection and identification, and an obstacle detection module. The tests revealed high performance of the tested modules.

We conducted experimental tests on path planning and satellite navigation trajectory tracking modules to evaluate the positioning ability of the system. The system performance is optimal in urban environment without tunnels or any other long-time disruptions of GNSS satellite connections. Our next step to increase the positioning accuracy is collaboration with diversification systems (vision system, laser scanning, etc.). A better quality of inertial unit output trajectory is needed as well. After all the necessary engineering follow-up more experimental test will take place to evaluate the corrections made.

This autonomous vehicle driving control system and the groundwork laid during the engineering development will give an impetus to self-driving public buses development. Autonomous public transport will cut transportation and time costs, increase economic efficiency and minimize road traffic accidents.

Acknowledgements

This research done with the financial support from Ministry of Education and Science of the Russian Federation (the unique project identifier is RFMEFI57717X0268).

The experimental research conducted with the use of measurement equipment of the NNSTU Centre of collective using «Transport Systems».

References

[1] M. Bojarski and et al., End to end learning for self-driving cars, 2016.
[2] Chen, C., and et al., learning affordance for direct perception in autonomous driving. In: ICCV.
[3] Krizhevsky and et al., Imagenet classification with deep convolutional neural networks. In Advances in Neural Information Processing Systems 25, 2012, pages 1106–1114.
[4] C. Hane, and et al., 3D Visual Perception for Self-Driving Cars using a Multi-Camera System: Calibration, Mapping, Localization, and Obstacle Detection, 2017, ArXiv eprints;
[5] J. Zhang, and et al., «LOAM: Lidar odometry and mapping in real-time», Robotics: Science and Systems Conference (RSS), 2014, July.
[6] Autonomous car development company Waymo. Retrieved May 30, 2018, from: https://waymo.com/
[7] The Official NVIDIA Blog. Retrieved May 30, 2018, from: https://blogs.nvidia.com/
[8] Baidu. Retrieved May 30, 2018, from: http://apollo.auto/
[9] DARPA. Retrieved May 30, 2018, from: https://www.darpa.mil/
[10] Oshkosh Defense. Retrieved May 30, 2018, from:https://oshkoshdefense.com/
[11] Lockheed Martin. Retrieved May 30, 2018, from:https://www.lockheedmartin.com/en-us/index.html
[12] SAE International. Retrieved May 30, 2018, from: https://www.sae.org/
[13] Yandex Taxi Unveils Self-Driving Car Project. Retrieved May 30, 2018, from: https://yandex.com/company/blog/yandex-taxi-unveils-self-driving-car-project/
[14] Cognitive Technology. Retrieved May 1300, 2018, from: https://www.cognitive.ru
[15] Central scientific research automobile and automotive engine institute «NAMI». Retrieved May 30, 2018, from: http://nami.ru/
[16] KAMAZ PTC. Retrieved May 30, 2018, from: https://kamaz.ru/
[17] Avrora Robotics. Retrieved May 30, 2018, from: http://avrora-robotics.ru/
[18] YOLO. Retrieved May 30, 2018, from: https://pjreddie.com/darknet/yolo/
[19] Robot Operating System. Retrieved May 30, 2018, from: http://www.ros.org/
[20] Basler the power of sight. Retrieved May 30, 2018, from: https://www.baslerweb.com/ru
[21] Pergam. Retrieved May 30, 2018, from: https://www.pergam.ru/
[22] Xsens. Retrieved May 30, 2018, from: https://www.xsens.com/
[23] OpenCV. Retrieved May 30, 2018, from: https://opencv.org/

IOP Conf. Series: Materials Science and Engineering **386** (2018) 012004 doi:10.1088/1757-899X/386/1/012004

Critical analysis of new equation of the wheel machine motion

V I Peskov, N A Kuzmin

Nizhny Novgorod State Technical University n.a. R.E. Alekseev, 24 Minin Street, Nizhny Novgorod, Russia

Abstract. The article is concerned with the discussion between the author of a new equation of the wheel machine motion who is the doctor of technical science, professor G I Mamiti and his opponents. The discussion continues over a year. The opponents note that the initial hypothesis taken as the basis at development of the new equation is incorrect and unreasonable by consideration of process of reduction of forces influencing a car body to contact of driving wheels with the road. As a result, large mistakes can take place when using this equation for calculations of dynamic properties of the cars. Four doctors and three candidates of technical sciences have expressed negatively about the new equation. Nobody in the central technical press has expressed positively. Therefore, it has been offered to finish a discussion currently.

Technical magazine «Automobile Industry» (Russia) organized the discussion about of the new equation of wheel machine motion the author of which is professor Mamiti G I from Gorsky Agrotechnical University (Vladicaucasus). The new equation is based on use of a wrong hypothesis of transfer to driving wheels of the moments which affects the case in the longitudinal direction from resistance forces the movement on rise, air resistance, from forces of inertia and forces from a trailer. Classical equation of automobile motion without of driving wheels sliding is given below [1]

$$F_{т} - \delta m_a a_x - G_a(f\bullet\cos\alpha + \sin\alpha) - k_w A_m V^2 - F_{cx} = 0, \qquad (1)$$

where $F_{т}$ – drag force on driving wheels; δ – calculation coefficient of revolving mass; m_a – mass of automobile; a_x – acceleration of automobile direct motion; G_a – weight of automobile; f – coefficient of rolling resistant; α – angle of automobile motion on ascent; k_w – coefficient of air resistant; A_m – middle square of automobile; V – speed of air motion relative of automobile ; F_{cx} – horizontal part of force on tow.

The original choice of the new equation of balance of force which was used by the author for measurement of process of transformation of force influencing a car body to contact of driving wheels with the road causes special coefficients of type $(h - r)/r$ (2) [2, 3, 4], in which h_g – height of position from surface of road of the automobile center mass; h_w – accordance height of position point of sum air resistant; h_c – height of position action point horizontal part of force on tow; r – wheel radius:

$$F_{т} - \delta m_a a_x(h_g - r)/r - G_a f\bullet\cos\alpha - G_a\sin\alpha(h_g - r)/r - k_w A_m V^2 (h_w - r)/r - F_{cx} (h_c - r)/r = 0, \qquad (2)$$

But even simple researches of this equation lead to the conclusion that any force which we can use doesn't show a resistance of the machine motion, if to do $h = r$ (or $r = h$), because in this case corresponding coefficient becomes equal to zero. It is clear, it is impossible in real life, so in his later mathematical works G I Mamiti [5] was forced to transform this equation by replacing of showed coefficients $(h - r)/r$ on h / r.

However there are some new difficulties: because almost every real wheel machine has $h > r$. All forces of resistance in new equation turn out to increase essential in comparison with classical equation. For example, for the bus with wagon body, which usually has high situation center of mass h_g and sail center h_w ($h_g \approx h_w$ = 1,3-1,6 m) and typical wheel radius 0,5 m coefficient of increase will be 2,6-3,2. As a result, according to the new theory of the wheel car, moving maximum engine power for achievement of necessary maximum speed will also increase by 2-3 times. According to the

IOP Conf. Series: Materials Science and Engineering **386** (2018) 012004 doi:10.1088/1757-899X/386/1/012004

classical theory, the rated size of the maximum engine power will be well-coordinated with the maximum engine power of the real tire.

Understanding that this essential lack of its new work seriously undermine the author's authority , G I, Mamiti corrected his new equation for the third time: he implemented the method of longitudinal transformation of forces only to force of air resistance without any reason. So, coefficient h/r in new variant of equation remained only at force of air resistance:

$$F_т - \delta m_a a_x - G_a(f{\bullet}\cos\alpha + \sin\alpha) - k_w A_m V^2(h_w/r) - F_{cx} = 0. \quad (3)$$

To prove correctness of such decision he referred to practice of construction of a race car [5, 6]. Really, for low sitting machines as famous «Formula-1» new equation gives minimum mistake in calculations in comparison with more correct calculations with the help of classical equation. The matter is that for machines of this class it is near real $h_w \approx r$ and if to take into consideration this equality in third generation of a new equation of motion, we will receive the classical equation (1) of force balance (3), which is used very widely in calculations of wheel machines dynamic in every days engineer practice. Influence of a high-rise action pointed of equal force of air longitudinal resistance will be fiction because there is no parameter h_w in that equation.

Now it is clear, that there is no sense in new equation. Especially, because it does not give correct calculation results for usual machines, when mistakes can reach 100-150% [1]. The cause of error consists in a wrong basic hypothesis about the transfer moments to driving wheels from action on mass resistance forces the movement on rise, air resistances, from forces of inertia and forces to leave a trailer. Nobody used the procedure called by the author as a new equation «leading of resistance forces to contact the driving wheels with the road». Transformation of one equation to other (as their author demonstrates in article [6]) can not be the proof, that a new equation of wheel machine motion is absolutely correct. It is because that both of these equations are created with use of a disputable method of forces of leaders.

In attempt to justify use hypothesis of transfer to driving wheels of action for the body moments from resistance forces, the author of the new equation addresses the fact that "... Automobile is a mechanical system with the different nature of separate elements connections and difficult laws of their relative movements. It is possible to ignore some relative movements of elements while different tasks decision " [6]. Due to this G I Mamiti makes the conclusion that the car can be considered as a rigid body for the analysis of the car dynamic traffic.

It is very difficult to believe, that the words belong to a doctor of technical science. After all it is impossible to use a rigid model of automobile for all cases. For example, it may be used for analysis of extra braking regime with blocking driving wheels. When automobile moves and wheels with transmission elements rotate, the monolithic rigid body model does not correspond to reality, so the conclusions about its behavior with the point of view of the dynamic movement will be wrong. It is impossible to use such model for consideration of characteristics of productivity of the engine, characteristics of creation of gear and gear ratios, inertia of moments of rotation of parts and others.

For the analysis of a situation, G I Mamiti refers to a similar example when a sail on the boat get down in case of wind strengthening [6]. It reduces the force of the wind, and it happens, first of all, because of reducing sail area, but not because of reducing the height of an action point equal to force of longitudinal air resistance. For example, multi-storey sails aren't got down on big sailboats. They are attached to a cross-piece mast. It means that the center of a sail rises, but the surface of contact with an air reduces. The main goal of this process is to reduce overturn of a boat. When we estimate an influencing factor of height of an action point equal to force of air, we mean air resistance to the movement of the car. The opponents of the new equation understand a way of creation of this moment and don't deny its existence, unlike G I Mamiti's statements. In relation to the car, at the time of a body moves to the longitudinal direction, vertical reactions of the road in front and behind changes, but it doesn't transfer to driving wheels. Driving wheels receive only a longitudinal part of force of longitudinal air resistance as counteraction.

Mamiti's attempts to adduce arguments in his favor as a constant on "publishing misfortune" -misprint aren't really convincing. In all conditions coefficients of this kind as (h – r) / r leading to revolutionary conclusions in the theory of the car about a possibility of neutralization of resistance forces in initial version of a statement about novelty of the equation of the movement which appeared consistently in Russia in the third different technical journals [2, 3, 4] can't just be typos. It means a gross blunder of the author, but he hasn't recognized this mistake yet.

It is also necessary to note that the longitudinal forces applied to a car body during attempt to start the car "hands" because of the discharged accumulator aren't transferred via the spherical hinge as it is dictated by G I Mamiti [6]. These forces transfer to axes of wheels from the parts of the suspender and bearings fixed on them connected to a body. So the corresponding longitudinal reactions of resistance to the movement in contacts of driving wheels with the road which in result (together with effort of pressing a wheel axle) create on them turning points are formatted, given by means of the spherical hinge and axial shaft to transmission and further – on the crankshaft. If this moment is rather vigorous to turn a cranked shaft engine can be started, it is necessary to switch off coupling only in time and to press the brake pedal because the car can collide someone or something ahead of it.

If the hypothesis of G I Mamiti of transfer of action for the body moments on driving wheels is right, then their turn in the considered situation will depend on height of application of the pushing force to a body: the point of application, the transfer and the moment of the crankshaft more twisting is higher, the engine start is easier. But there is no such dependence, nobody noted it in operational practice, nobody advised in attempt to start the car "hands" to concern by force of pushing of the people to the highest point of the car. Special experiments confirmed that unlike application of longitudinal force to a body, application of the longitudinal moment leads to the movement of the car [1]. So the address to opponents "to go down on the earth" for start of their automobile engine in case of the discharged accumulator confirms, but destroys an initial hypothesis of the author of the new equation of the movement of the wheel car.

Everything told in this article explains the fact that four doctors and three Candidates of Technical Sciences from leading universities of Russia and Ukraine carried out the tough critical analysis of the new equation [1, 7, 8]. In the central technical publications there are no articles in support of its new theory of the movement of wheel cars. Therefore, it is time to complete this discussion which lasts more than a year.

List of reference literature

[1] Kuzmin N A, Serduk V I and Peskov V I 2016 *Analysis of new equation of wheel machine movement* (Moscow: Tractors and Agrimachines, №4) pp 24-28
[2] Mamiti G I 2014 *New equation of wheel machine movement* (Moscow: Tractors and Agrimachines, №6) pp 15-18
[3] Mamiti G I 2015 *New calculation force scheme of wheel machine and equations its movement* (Moscow: News of Machinebilding, №4) pp 35-37
[4] Mamiti G I 2015 *New equation of movement, force calculation, definition of middle square and its centre* (Moscow: Automobile Industry, №10) pp 16-18
[5] Mamiti G I 2017 *New equation of wheel machine movement right* (Moscow: Automobile Industry, №3) pp 22-27
[6] Mamiti G I 2018 *New equation of wheel machine movement and objection its opponents* (Moscow: Automobile Industry, №4) pp 25-28
[7] Kopotilov V I 2017 *Once more about new scheme of calculation and equation of wheel machine movement* (Moscow: Automobile Industry, № 7) pp 31-36
[8] Podzidaev S P 2018 *Triumph of science on sound sense* (Moscow: Automobile Industry, № 4) pp 20-25

IOP Conf. Series: Materials Science and Engineering **386** (2018) 012005 doi:10.1088/1757-899X/386/1/012005

Critical speed of propshaft and effect of forced synchroniztion in transmission of straight-four engine vehicles

V A Saveliev

PhD, FSUE "NAMI", Moscow

E-mail: cardan-nami@yandex.ru

Abstract. The article shows the role of precession of a propshaft in increasing its vibration ativity. An auto inline four-cylinder engine has been shown to trigger a resonance mode of its drive shaft vibrations, i.e. intense flexural vibrations, leading to an emergency situation. This is due to the effect of forced synchronization (frequency capture) when the frequency of the second order unbalanced inertia forces is approaching the critical frequency of drive shaft rotations.

1.Introduction

Unbalance degree reduction of the different machine rotors is the primary task for solving the problem of their increased vibration activity, for which reason the most accurate possible definition of their unbalance parameters is necessary. This concerns both rigid rotors operating at the speeds, which are lower than the critical ones, and flexible rotors, operating at the overcritical speeds [1, 2]. However, when solving the problem of unbalance parameters identification, calculation results are often inconsistent with the experimental data and 'slight errors in rotor vibration measurements lead to big errors in unbalance parameters definition' [3]. Unbalance of a rotating propshaft is also one of the main causes of the emergence of vibrations, affecting its supports and being transferred to the units (aggregates) connected by it and the vehicle design as a whole. Upon that, it is important to keep in mind that the rotating propshaft carries out complicated motion, which can be represented as a superposition of axial rotation and precession. Ways of solving the problem of reduction of the different machine rotors unbalance are well-known. Theory and physics of the phenomenon related to it, including those related to the precession motion, are sufficiently completely considered in the rotor dynamics literature [4 - 9]. In existing literature, quite insufficient attention is paid to the precession motion problem in relation to the propshafts of transmissions of automobiles and other vehicles. In this literature, the consideration of the rotation motion dynamics is limited by the critical speed calculation. Critical speed is an important parameter characterizing the behavior of the propshaft under the operating conditions, but its vibration activity performance is largely defined by the precession motion, rather than self-rotation, which is usually left out of discussion of calculation methods of the resonant mode parameters. What is important is not only defining the critical speed of the propshaft, but also exploring the nature of the problem and showing a movement from the phenomenon to the crux of the problem. Speed corresponding to the sharp icrease in the amplitude of vibrations excited by the propshaft unbalance is taken as a critical speed. This phenomenon is usually explained as a resonant increase in the vibration amplitude at the disturbing forces action frequency caused by the rotating shaft unbalance, which coincides with the own speed of the flat (planar) bending vibrations of the non-rotating shaft, which is logically inconsistent, whereas effects detected at the rotation motion are deduced from the conditions obtained for the non-rotating system. An approach to the explanation of the rotating shaft behavior from the perspective of the consideration of the time behavior of the small disturbances of its established motion, which is a rotation with a constant angular speed, seems more logical than from the perspective of the analysis of its bending vibrations in relation to static balance position. This more precisely reflects the phenomenon and is especially important in the analysis of the behavior of the considered object during the rotation at overcritical speeds [5]. For the effects analysis, that may appear at the propshaft rotation, let us consider two cases: a) an ideal case, which is the rotation of a balanced propshaft; b) rotation of an unbalanced propshaft.

2. Rotation of a balanced propshaft

The rotating rotor will not cause the alternating disturbing influence on its supports, if the rotation axis coincides with one of its main axes of inertia [1, 2]. Upon that, a mass fixed on a rigid simple (two-point) beam is traditionally taken as a calculation model of such structures, i.e. the rotor rotation axis deflection caused by weight load is ignored. In order to ignore this phenomenon, in certain studies the rotor rotation is considered at its vertical configuration. The rotor model with a vertical configuration of the rotation axis is applied in order to simplify the solution of its dynamics problems [8, 10]. However, the propshaft is not usually vertically installed in vehicle transmission and has a deflection caused by the action of weight load. Upon that, the propshaft rotates about the obtained bending line. As the propshaft mass center and the centers of sections of its elementary weights are on the bending line, the balanced two-point propshaft with a ring cross-section pipe shall not cause vibrations at self-rotation.

3.Rotation of an unbalanced propshaft

During rotation of an unbalanced propshaft, there is a centrifugal force $R = m\omega_{sq}^2 r$, where m is the unbalanced mass; r is the radius of the location of the unbalanced mass; ω_{sq} is the propshaft speed. The centrifugal force and weight load during rotation of an elastic propshaft, i.e. a structure having the finite stiffness in a plane of bending, result in change of its kinematic behavior. And upon that, rotation of the propshaft around the bending line caused by the weight load with angular speed ω_{sq} is supplemented by precessional motion – rotation of the propshaft bending plane with angular speed Ω, which is called precession rate. Such conclusion is consistent with the results of analysis of precession motion of a rotor in a form of massless shaft with a disk fitted on it, while 'the precession axis turns out to be not the bearing axis, but a curve defining the shaft position under static (in this case – weight) load' [9]. Thus, the weight load and unbalance of the propshaft cause precession of its axis and, consequently, vibrations transmitted to the attached units (aggregates) through the supports of the rotating shaft. The first works (1869) on rotor dynamics connected with the 'rotor precession' phenomenon refer to the name of W. Rankine, who called it 'centrifugal whirling' [8]. A pin-supported rotating propshaft is some kind of gyro system that reminds a rate gyroscope. One of the degrees of freedom of such gyro is formed by its own rotation around the curved axis with angular speed ω_{gyr}, and the other – by rotation of the bending plane with angular speed Ω. In mechanics, a gyroscope is a rotating solid body of any shape; upon that, its speed is not a characteristic feature [11]. The problem of motion of such body 'in general is still not solved and, in a certain sense, nonsolvable' [6]. To predict the behavior of gyro systems the conservation of angular (impulse) momentum law – the fundamental law of physics – is used, according to which

$$\frac{dL}{dt} = M, \qquad (1)$$

where $L = \omega_{gyr} I$ is the impulse moment (kinetic moment or angular moment), I is the moment of inertia of a rotating body. If moment M of external forces applied to the gyroscope equals zero, then angular moment $L = const$, and the gyroscope rotation axis will maintain certain position in space. If the moment of external forces is not zero, then the gyroscope will move while one of its points will remain fixed. In this case, the gyroscope motion can be considered as rotation around an instantaneous axis that goes through a fixed point. For the rotating propshaft, such points can be its pin-bearing supports. The interrelation of angular speeds of precession Ω, rotation about own (figure) axis ω_{gyr}, external moment M and moment of inertia of a rotating body I according to the approximate theory of gyro systems is expressed as follows:

$$\Omega = \frac{M}{\omega_{gyr} I} \quad . \tag{2}$$

It should be noticed, that the precession of a rotating shaft can be both direct and reverse depending on the direction of angular speeds of the shaft itself and its precessional motion, which are defined, for example, by spacial characteristics of elasticity of the shaft supports, including their anisotropy or isotropy [13, 14]. Upon that, in general, precession frequency of a rotating shaft Ω depending on speed ω_{gyr} of the shaft itself is defined through coefficient s which can be an integer or a fraction, positive or negative, i.e. $\Omega = s\omega_{gyr}$ [7]. However, inertia forces of unbalanced masses of the most common rotor structures, including vehicle propshafts, generate forced vibrations of these structures in a form of synchronous precession with a frequency equal to their speed [2, 7] and for them $s = 1$ and $\Omega = \omega_{gyr}$. These forces are proportional to the square of the rotor speed. The source of disturbance can be the rotor weight and any other unidirectional load as well. In the educational literature, motion analysis of statically unbalanced rotors is often limited to the consideration of direct synchronous precession. In [9] the physics and conditions of excitation of direct (forward) and backward rotor precession are considered. The backward recession of unsupported rotating propshaft in certain conditions of its excitation has been registered during tests by means of photodiode sensors [15].

4. Synchronization of the frequency of action of unbalanced second order inertia forces of the engine and the critical speed of the propshaft

The area of dangerous self-excitation of bending vibrations of the propshafts can be rather precisely defined within the procedure, whose description is provided in educational literature [16, 17], in standard [18]. However, solution of this problem is not exhaustive. For example, such a solution cannot be called sufficient for the case of in-vehicle application of widespread in-line four-cylinder engines. It is connected with the fact that, as known [19], crankshaft rotation in operation of the in-line four-cylinder four-cycle engine with angular speed ω is followed by the action of unbalanced second order inertia forces with frequency 2ω. These forces are transferred to the unit (aggregate) supports (mounts) and the elements of dynamic system 'engine crankshaft – clutch – gearbox – propshaft' elastically attached to the flexible frame, i.e. with the common base. At coincidence of frequency $2\ \omega$ of action of unbalanced second order inertia forces of the engine and the critical speed of the propshaft ω_{cr}, the latter enters the mode of resonance of the bending vibrations characterized by the sharp increase in amplitudes of the bending vibrations. Thus, if frequency $2\ \omega$ of action of unbalanced second order inertia forces on the engine mounts approaches the critical speed of bending vibrations of the propshaft ω_{cr} connected via the gearbox and the clutch with the crankshaft rotating with frequency ω, there is forced synchronization of vibrations, frequency capture. Upon that, the propshaft is exposed to intensive bending vibrations with frequency $2\ \omega$ of the synchronizing system, as it begins to operate in the resonance area that can result in the destruction of the propshaft [20-22]. Therefore, the critical frequency (speed) ω^{sync} corresponding to the forced synchronization of vibrations can be determined as follows:

$$\omega = \omega^{sync} = 0{,}5\omega_{cr} . \tag{3}$$

Here ω_{cr} - the propshaft critical speed (according to the first form).

5. Figure

Figure 1 shows the scheme of action of unbalanced second order inertia forces of to-and-fro moving masses of the engine on its mounts. This figure shows as follows: 1 – engine, 2 – an elastic element of engine mounting (suspension), 3 – basis of the engine mount, 4 – a rotation axis of unbalanced mass, 5 – the vibrator housing, 6 – a core, 7 – vibrator unbalanced mass. The vibrator is schematically depicted on the figure in the form of two equal unbalanced masses symmetrically located in relation to the vertical axis and rotating in opposite directions (in antiphase) with frequency 2 ω. Upon that, the horizontal component of centrifugal forces at every instant is equal to zero, and only the vertical component of centrifugal forces is transferred to the mounts [23].

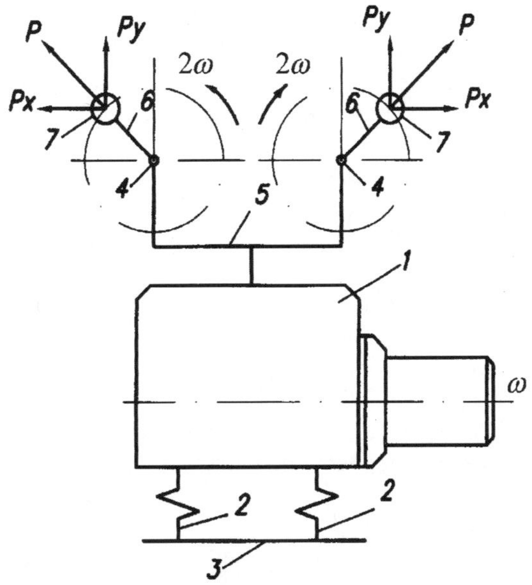

Figure 1 The scheme of effect of unbalanced second order inertia forces of to-and-fro moving masses of the engine on its mounts

6. Conclusion

6.1 The propshaft is one of the significant sources of the vibrations transmitted from the transmission to the vehicle design as a whole. In the process of its operation, it carries out complicated motion, which can be represented as a superposition of its axial rotation and precession.

6.2 The current state of the rotor dynamic processes theory allows making a conclusion that a rather precise reflection of the physics of increase in vibration activity of the propshaft rotating in the vehicle transmission is possible from the perspective of its precession motion analysis.

6.3 In case of in-vehicle installation of in-line four-cylinder engines it is evident that the effect of forced synchronization (frequency capture) may lead to the following: if the frequency of action of the unbalanced second order inertia forces of the engine approaches the critical propshaft speed, these propshafts will enter the resonant vibration mode and will be exposed to the intensive bending vibrations making an emergency possible. Upon that, to ensure safety during operation under the operating conditions, determination of the critical propshaft speed margin is necessary but insufficient. In this case, the additional condition is taking into account the oscillating system of the unbalanced second order inertia forces of the engine.

7. References

[1] Gusarov A A 1990 Dynamics and balancing of flexible rotors (Moscow: Nauka), p152

[2] Nikiforov A N State of balancing rotors problems 2013 *Journal of Scientific and Technological Development* 4(68) pp. 20-28.

[3]Lyakhov A F 2017 Computational Conditionality of the rotor balancing problem *Lobachevsky State University of Nizhni Novgorod* (Nizhni Novgorod, Russian Federation) **79** 2 pp 135-146

[4] Skubachevskiy G S 1969 Aircraft gas-turbine engines. Design and calculation of parts. 3rd edition, updated and revised (Moscow: Mashinostroenie Publ.) p544

[5] Grigorev N V et al 1974 Vibration of power units. Handbook. Ed. Grigorev N V (Leningrad: Mashinostroenie Publ.) p464

[6] Arnold V I 1979 Mathematical methods of classical mechanics (Moscow: Nauka Publ., Glavnaya redaktsiya fiziko-matematicheskoy literatury) p432
[7] Khronin D V 1980 Vibrations in aircraft engines 2nd ed. (Moscow: Mashinostroenie Publ.) p 296
[8] Pasynkova I A 2014 Dynamics of precession motion of unbalanced rotor (Saint-Petersburg: Izdatelstvo Sankt-Peterburgskogo Universiteta Publ.) p238
[9] Podolskiy M E, Cherenkova S V 2014 Physical nature and conditions of initiation of direct and reverse precession of rotor *Teoriya mekhanismov i mashin* vol. **12** 1 pp 27–40
[10] Mitenkov F M et al 2007 On stationary movements of the distributed and discrete models of a rotor on an electromagnetic bearing *Lobachevsky State University of Nizhni Novgorod* (Nizhni Novgorod) 1pp 130-136
[11] Magnus K 1974 Gyroscope. Theory and applications (Moscow: Mir Publ. Translated from German by Blyumin G D et al) p526
[12] Landau L D, Lifshits E M 1988 Theoretical Physics Vol **1** Mechanics (Moscow: Nauka Publ.) p214
[13] Ryzhik B V 1986 The influence of unbalance on the fatigue strength of the rotor supported by anisotropic elastic-damping supports *Mashinovedenie* (Moscow: Nauka Publ.) 3 pp 34–38
[14] Pasynkova I A 2014 Forced oscillations of an unbalanced rotor in nonisotropic bearings *Vestnik SpbGU* Ser. 1 Vol. **1**(59) Issue 2 pp 292–302
[15] Denisov G G et al 1969 Experimental research of unsupported rotating propshaft oscillations *Vehicle dynamics Collect.* (Moscow: Mashinostroenie Publ.) pp 127–138
[16] Lukin PP, Gasparyants G A, Rodionov V F 1984 Vehicle design and calculation (Moscow: Mashinostroenie Publ.) p 376
[17] Design of vehicle transmissions. Guide. Ed. Grishkevich A I (Moscow: Mashinostroenie Publ.) 1984 p272
[18] GOST 52430-2005 Vehicles. Cardan gears of vehicles with joints of unequal angle speeds. General technical requirements (Moscow: Standartinform Publ.) 2006 p 12
[19] Tolskiy V E et al 1976 Vibrations of vehicle powertrain (Moscow: Mashinostroenie Publ.) p 266
[20] Ivanov S N, Saveliev V A, Kocheshkov N P 2010 Aspects of calculation of cardan drive in order to ensure its safe work during operation *Zhurnal AAI* 5 pp 46–48
[21] Saveliev V A 2010 The features of the influence of unbalance of rotating masses of vehicle transmission on its vibration activity *Safety of vehicles in operation: proceedings of the 71 International Scientific and Technical Conference* (Nizhny Novgorod, October 12-13) pp. 20–22
[22] Saveliev V A 2012 Influence of unbalanced forces of inertia of the second order of the four-cylinder row engine on flexural oscillations of cardan drive shafts of a vehicle *Vestnik of Lobachevsky University of Nizhni Novgorod* 5 (2) pp 197–200
[23] Litvak V I 1972 Automation of fatigue tests on full-scale structures (Moscow: Mashinostroenie Publ.) p 384

IOP Conf. Series: Materials Science and Engineering **386** (2018) 012006 doi:10.1088/1757-899X/386/1/012006

Determining friction coefficient between the base cylinder of rotary screw propulsion unit and various ice types in different daylight surface

I A Erasov[1], Yu I Molev[1], D S Mokerov[2], T G Shetulova[1]

[1]Nizhny Novgorod State Technical university after R.E. Alekseev, Russia
[2] Kozma Minin Nizhny Novgorod State Pedagogical University, post-graduate, Russia
E-mail: erasoff2013@yandex.ru[1], moleff@yandex.ru[2], mokerov90@bk.ru[3]

Abstract: Statistical data processing results of friction coefficient change between steel base cylinder of rotary screw propulsion unit and ice of different surface roughness are considered in the article. Experimental values of ice surface roughness complex criterion are given. It is also established that in actual operating conditions the basic factor affecting the friction coefficient change of the base cylinder on ice is represented by the processes connected with the plastic yield of the ice daylight surface irregularities. At this the adhesive component of friction force while operating on rough ice did not exceed the 20% value.

To improve social and economic standards and to carry out projects of transport infrastructure development in the Arctic zone, it is necessary to create new as well as modernize current all-terrain vehicles as far as underpopulation and thus low cargo traffic makes the road network development in this region unprofitable.

Nowadays over 100 kinds of special-purpose snow and swamp-going vehicles are produced worldwide, capable of performing transportation and processing operations in swam and snowscape area. Herewith, in order to cut the production costs most of these vehicles are manufactured on the basis of production cars. Despite the fact that these vehicles passed certification tests during production, including the ones measuring the amount of noise in the vehicle, the noise impact on the driver in operating all-terrain vehicles created on their basis, frequently exceeds the established norms and can reach 90 to 100 dB. Thus we can make a conclusion that all-trailer propulsion unit interaction with the supporting medium is the source of excessive noise when snow and swamp-going vehicles are in motion. Consequently studying the interaction of different propulsion unit types with the supporting medium in order to decrease the noise level in the vehicle and operating medium is an important scientific problem.

It should be noted that unlike cars, all-terrain vehicles move over surfaces having high diversity of physical and mechanical, strength and geometry parameters. Hence the most rational approach to simulate the noise level is to develop several models of its appearance for each propulsion unit type interaction with the given supporting medium. The article considers one of the most widespread types of motion – the interaction of rotary screw propulsion unit with ice. In this case the noise level will be determined solely by the results of friction of the base cylinder and screw blade on the ice surface as shown in Fig.1.

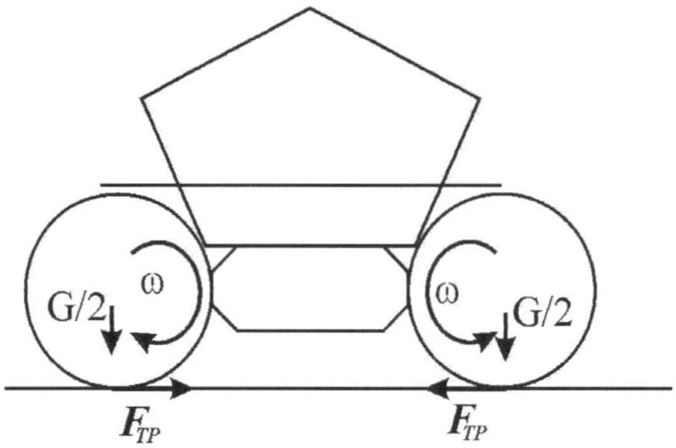

Figure 1. The chart determining the noise level created by the base cylinder at its friction on ice.

The noise when the rotary-screw machine is in motion appears due to the interaction of the rotor and the supporting medium and depends on the contact surfaces motion speed. This noise is caused by the base cylinder hitting the ice microroughness. Friction force on each peak of contact surfaces according to [1-2] can be presented by 2 summands: a molecular (adhesive) and a mechanical component. Thus the total friction force depends on the contact surfaces actual field of contact. The surface roughness is characterized by the complex roughness index Δ. At its low values, i.e. at a high class of surface smoothness, the dominant factor is a molecular component; and at increasing Δ in the contact zone plastic yield prevails [6]. Determining the value of the complex index Δ is a complicated applied research task which involves determining both the ice surface roughness height and the radius at their asperity summits. This is possible only by statistical processing of test data carried out within the framework of federal task program «Research and development in priority development fields of Russia's science and technology sector for the period of 2014-2020 » № 14.577.21.0222 its subject being «Creating a test sample of amphibious stand-alone transport and technological complex with intelligent control and navigation system for carrying out all-year-round exploratory drilling operations on the Arctic shelf». According to the research the dependency of Δ index and the maximum surface roughness height can be approximated by the dependency:

$$\Delta = 0{,}038 h_{MAX} - 0{,}05$$

For the circuit load, equal to 1 rotor load rate ratio to the rotor area of contact with ice surface, equal to the ice compression strength HB 3-6 MPa [3], elasticity modulus E, ranging from 4 to 8 MPa, and Poisson's ratio of 0,31 the contact index value of rotary screw propulsion unit equal to:

$$\frac{p_c(1-\mu)}{E} = \frac{(3 \div 6)\cdot 0{,}68}{4 \div 8} = 0{,}50 \div 51$$

can be obtained.

As soon as this index exceeds the contact index value of the rotary screw propulsion unit with the supporting medium equal to $0{,}06\cdot\sqrt{\Delta}$ [7] or 0,1, it is possible to make a conclusion that the contact of the rotary screw propulsion unit with ice is of intense yielded type. For this interaction type the friction coefficient value of these surfaces is calculated by the equation [7]:

$$f = \frac{\tau_0}{HB_C} + \beta + 0{,}9\sqrt{\Delta\frac{HB_{Л}}{HB_C}}$$

where τ_0 — is the observed tangential strength at the adhesive bond section; β — the observed pressure-coefficient of molecular friction component, $HB_{Л}$ and HB_C – are ice and steel indices in MPa respectively. In the case of ice plastic contact with steel, index τ_0 is determined by value 10-15 MPa, and β=0,0072. Then the friction coefficient of the steel base cylinder on the ice surface is determined by the expression:

$$f = \frac{\tau_0}{HB} + \beta + 0{,}9\sqrt{\Delta\frac{HB_Л}{HB_C}} = \frac{10}{2500} + 0{,}0072 + 0{,}01\sqrt{0{,}038h_{MAX} - 0{,}05}$$

$$f = 0{,}0112 + 0{,}01\sqrt{0{,}038h_{MAX} - 0{,}05}$$

Thus, the dependency which considers not only steel characteristics and ice temperature data, but also ice roughness, while determining steel friction coefficient upon ice, was obtained.

The value of correlation coefficient with the test data is 0,996. The data obtained are shown in Fig.2.

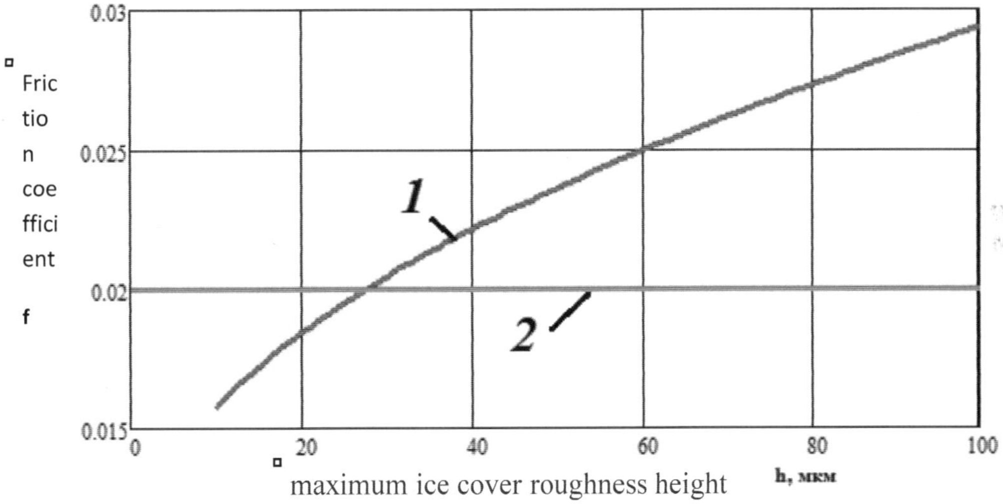

Figure.2. Calculation data of steel on the ice friction coefficient considering. roughness parameters of ice; 1 – dependency obtained; 2 – previously applied dependency

Calculation data show that the calculation method used up to now had a calculation error of over 60% when describing the machine operation on rough ice as it did not consider ice surface roughness parameter. The research shows that this parameter is crucial. The newly obtained dependency allows to improve calculation accuracy when determining both motion resistance force and noise level, created by the rotary screw propulsion unit.

References

[1] Berkovich I. Tribology. Physical basis, mechanics and engineering applications: College textbook / I. Berkovich, D. Gromakovsky; Edited by D.G. Gromakovsky – Samara State Technical University. Samara, 2000. – 268 pp.

[2] Fundamentals of tribology (friction, deterioration, lubrication) / Braun E., Bushe N., Buyanovsky I. et.al./ Edited by Chichinadze A.: Technical College textbook. M.: Center „Nauka i technika“, 1995. 778 pp.

[3] Makarov, V., Zeziulin, D., Belyakov, V Prediction of all-terrain vehicles mobility in snowscape scenes. //18th International Conference of the International Society for Terrain-Vehicle Systems, ISTVS 2014; Convention Center, Hoam Faculty House, Seoul National University, Seoul; South Korea; 22 September 2014 through 25 September 2014; Code 111443

[4] Belyakov V, Kurkin A, Makarov V, Zeziuli D Multifunctional vehicle for coastal areas // The Twelfth International Conference on the Mediterranean Coastal Environment (MEDCOAST 2015) 06-10 October 2015, Varna, Bulgaria p.945-951 ISBN 978-605-85652-3-4 (ISBN 978-605-85652-4-2)

[5] Belyakov V, Kurkin A, Makarov V, Zeziuli D Minaev D, Zaytsev A, Teslenko D Ground vehicle for ice conditions monitoring// Proceedings of the Thirteenth International MEDCOAST Congress on Coastal and Marine Sciences, Engineering, Management and Conservation, MEDCOAST 2017, 31 October – 4 November 2017, 2017, Mellieha, Malta, E. Ozhan (Editor) p.775-785 ISBN 978-605-85652-6-5 (ISBN 978-605-85652-8-9)

[6] Kombalov V. Impact of solid body roughness on friction and deterioration. M: Nauka, 1984. 110 pp.

[7] Keropyan A. Determining rational values of friction coefficients and roughness of rail and locomotive tire working surfaces // Thesis for a Doctor of Engineering degree in 05.05.06. «Mining machinery». – Ekaterinburg 2015. – 233 pp.

IOP Conf. Series: Materials Science and Engineering **386** (2018) 012007 doi:10.1088/1757-899X/386/1/012007

Development approach to a method for monitoring of driver's ability of resumption of control over the vehicle by on-board systems in automatic mode

V V Savchenko[1], S N Poddubko[2]

[1]Chief of R&D Center "Onboard control systems of mobile machines", The State Scientific Institution "The Joint Institute of Mechanical Engineering of the National Academy of Sciences of Belarus", Minsk, Belarus
[2]Director General, The State Scientific Institution "The Joint Institute of Mechanical Engineering of the National Academy of Sciences of Belarus", Minsk, Belarus

[1]uus@tut.by
[2]bats@ncpmm.bas-net.by

Abstract. The structure and basic functionalities of the upper level on-board information-analytical system and its link-up while simultaneous use with other intelligent on-board systems on the vehicle's board are studied. A development approach to a new method for estimating the potential for the transition (resumption) of driver's control over the vehicle by on-board systems in automatic mode, presumably having high efficiency and reliability, is proposed and justified.
Key words: ADAS, on-board intelligent systems, detection of hazardous driving conditions, vehicle, driver's functional state, vehicle automation level, resumption of driver's control over the vehicle.

Introduction

The widely known classification system of Levels of Driving Automation for On-Road Vehicles was developed by the Society of Automotive Engineers (SAE J3016 standard [1]). It contains 6 levels:

Level 0 (No Automation): no automatic control over the vehicle, but a warning system can be available; the full-time driving performance is carried out by the human driver.

Level 1 (Driver Assistance): the human driver mostly performs the driving; as a rule, Advanced Driver Assistance Systems are installed on the vehicle's board assisting the driver in steering and braking systems control using the information about the driving environment. They work in automatic mode, but can be disabled by the human driver.

Level 2 (Partial Automation): the human driver must react if the system fails to perform in the automatic mode. ADAS simultaneously controls acceleration, braking and steering, but can be disabled by the driver.

Level 3 (Conditional Automation): the human driver may not control the vehicle on some types of roads (for example, autobahns), but should be ready to resume control when changing the traffic conditions or the system stops the vehicle if the driver does not respond to the alert.

Level 4 (High Automation): it is similar to the 3rd level, but does not require the driver's attention and performs its functions permissible in specific external driving conditions.

IOP Conf. Series: Materials Science and Engineering **386** (2018) 012007 doi:10.1088/1757-899X/386/1/012007

Level 5 (Full Automation): no human intervention is required except for indicating the start of the driving and determination of destination.

The classification presented shows that right from the Level 1, on-board systems are hardware-software tools where a significant place (including in functionality) have the methods and algorithms for analysis of heterogeneous information on the functioning processes of vehicles and synthesis of control actions on different hierarchical levels systems for vehicle stability control and vehicle trajectory movement using IT methods and technologies and specialized software (the degree of automation increases at each Level), and only at the Level 5 no action is required on the part of the human driver except for indicating the start of the driving and determining the destination.

Figure 1 shows the vision of Movement company [2] on the current software "volumes" (one of the simplest characteristics of the metrics is the lines of code number) in different systems, in comparison, and how the "todays"-car will be changed in the future.

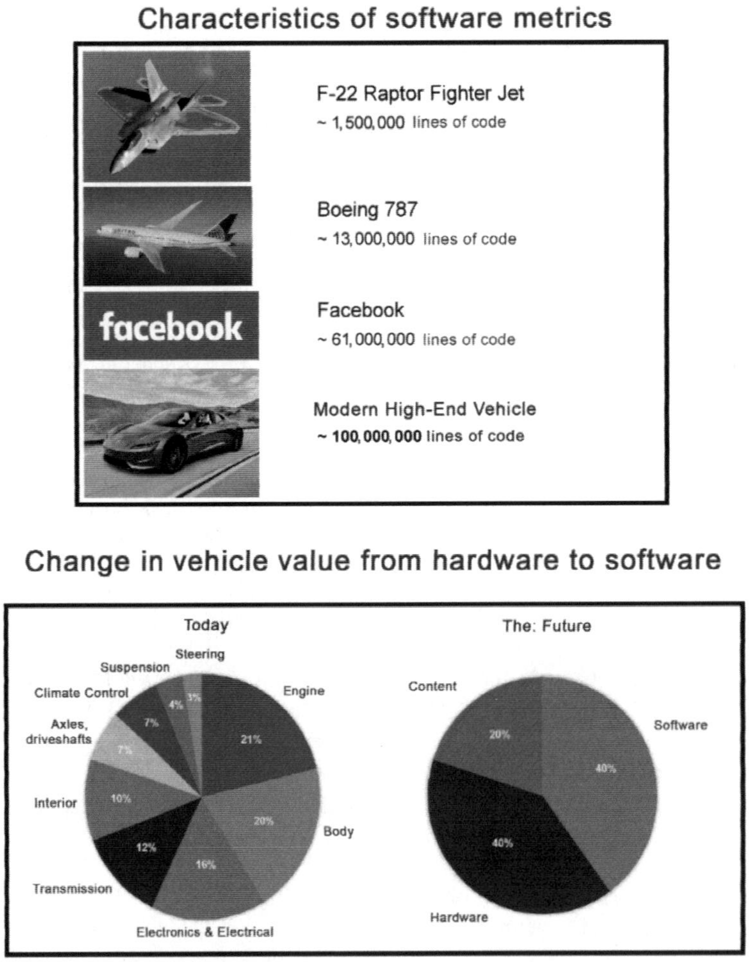

Figure 1. The number of lines of code in different systems and the volume of technical components, software and content in road transport in the future [2].

Thus, the level of automation of the vehicle significantly increases, while the stage of equipping a vehicle with ADAS systems is considered as an intermediate stage when moving to fully automated traffic mode. According to experts, by 2035 on the general use roads of the Russian Federation there will be about 10% of unmanned vehicles. Compromise options will be implemented, when the vehicle drives automatically under the specified conditions, and when the conditions change (primarily associated with the complexity or inability to formalize it) it "transfers" the control over the vehicle to the driver.

Classification of methods for driver fatigue monitoring

There are basic requirements for the systems of monitoring of the functional state of "human-machine" systems'operators [3,4], including vehicle drivers. One of these requirements is to identify the signs of hazardous states and to interpret the real-time functional states incompatible with the task performed (death, loss of consciousness, sleep, etc.). Detection of the hazardous states signs allows taking preventive measures in unmanned or automated mode, while statement of the presence of such functional states almost inevitably leads to emergency situation developing.

There are some methods for determining indications of sleep and deep relaxation of vehicle operators(Table 1, where p– theprobability of a hazardous failure, EDA – electrodermalactivity) to be the basis for many driver fatigue monitoring systems [5].

Table 1.Methods determining indications of sleep and deep relaxation [5].

Vigilance estimate technology	*p*
Changing the driving "style"	0,3
Rational actions	0,3
Pulpus	0,3
Posture (muscle tonus)	0,2
Glance direction	0,2
Head tilt (muscle tonus)	0,1
Speech	0,1
Oculography	0,05
Blinking	0,02
Microsaccades (optional)	0,001
EDA (72mln man-h without accidents)	0,0001

Development of upper-level information-analytical system

Fig. 2 demonstrates the structure of the upper level on-board information-analytical system (BIAS) and its link-up while simultaneous use with other intelligent on-board systems on the vehicle's board [6]. The main purpose of the system is preventive automatic prediction of an emergency situation, including those caused by erroneous driver's actions or automatic interpretation of driver's hazardous states (not compatible with the implementation of the basic algorithms for driving a vehicle), in real time, by keeping the vehicle in the traffic lane (with the use of active safety systems serially produced by OJSC "Ekran", systems of electromechanical power steering system for monitoring dead zones around a moving vehicle developed in Belarus, an electronic-pneumatic brake drive system, an adaptive cruise control system in the future), using classical passive ADAS potential (for example, Mobileye 6-Series), with adjustments, if necessary, of the trajectory movement of the vehicle using information from a high-precision navigation module. The BIAS fulfills the functions of the system integrator (depending on the on-board systems installed on a particular vehicle), reduces kinetic energy and/or stops the emergency mode of the vehicle if the driving conditions are appropriate, taking into account the integrated assessment on the basis of analysis of information from all the connected on-board systems from the unified methodological positions. In case of an accident occurrence, the BIAS implements automatically the function of emergency response systems in case of accidents (ERA-RB, ERA-GLONASS).

IOP Conf. Series: Materials Science and Engineering **386** (2018) 012007 doi:10.1088/1757-899X/386/1/012007

The on-board module for detection of driver's hazardous states is connected to the system via CAN bus [5, 6]. Application of the module significantly expands the functionality of the BIAS and gives the system the properties of a preventive safety system that secure from the occurrence of an accident because of deep relaxation, loss of consciousness or death of the driver. The module has some distinct advantages while it has a specialized chipset, robust algorithms for extracting a useful signal from the noise, original algorithms for automated interpreting of the driver's functional state, near telemetry for transmitting information to the stationary part of the system, which makes it possible to obtain high reliability of interpreting the dynamics of the driver's vigilance level comparable to the requirements specified for active safety systems. The prototype for the module being under development is a serially produced driver vigilance telemetric control system or VIGITON, which uses monitoring and automatic analysis of the parameters of electro dermal activity (the main criterion for automatic interpretation of the current dynamics of the driver's relaxation level, the method used has the smallest known probability of a dangerous failure (p– 0.0001), the concepts of building the driver's state safe monitoring systems are applied. The on-board module technical specifications for detection of driver's hazardous states (deep relaxation, loss of consciousness, death) contain the dangerous error probability of the registration method (2nd kind error) of no more than 4 10^{-4} and the probability of dangerous failures leading to accidents of no less than 10^{-9}.

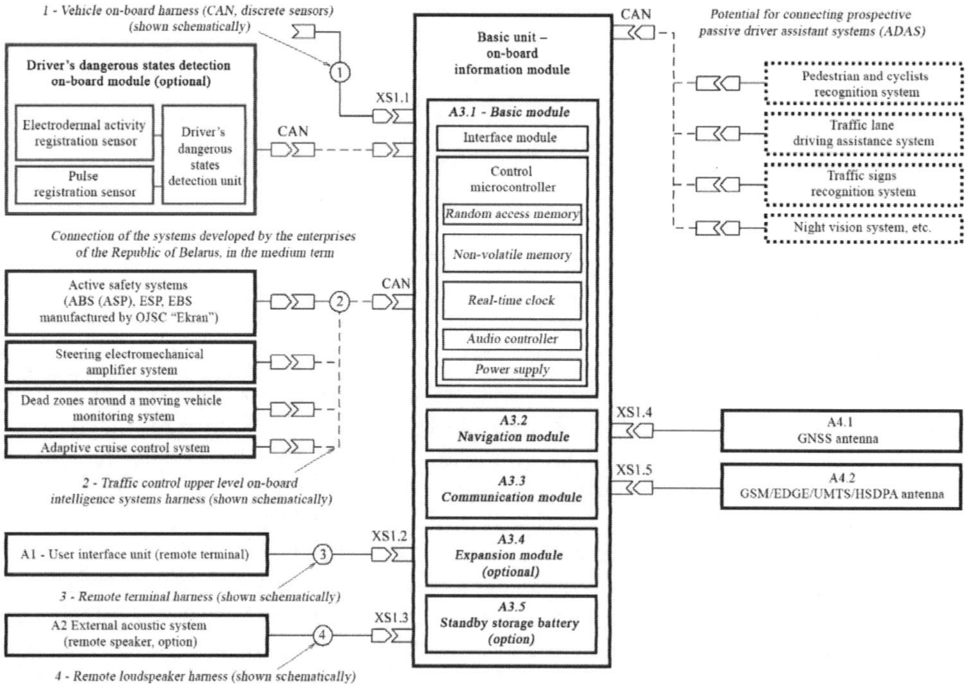

Figure 2. The BIAS structure and its link-up on the vehicle's board

The navigation module. GLONASS (L1, L2 frequency bands), GPS (L1, L2 frequency bands), GALILEO (E1B, E1C, E5a, E5b frequency bands) and BEIDOU (B1-C range) are used to improve the accuracy and reliability of the navigation module. The RTK technology (Real Time Kinematic) is provided for possible use for further improvement of position determination accuracy, which allows obtaining of centimeter accuracy of position determination due to the simultaneous processing of signals from GNSS and signals of correcting stations in real time [7]. The ability to integrate GNSS signal processing with the information from sensors of micro electromechanical systems and the odometer sensor of the vehicle's path allows using the "Dead reckoning" technology, which provides diminishing of the error in calculating of navigation parameters and allows position data assessment when the GNSS signals are inaccessible in the specified time spans. The module supports automatic monitoring of the integrity (accuracy) of navigation definitions and deletion of unreliable measurements, implementation of differential correction robust algorithms (RTK mode) and algorithms for integrating GNSS data and inertial sensors (accelerometers) of Belarusian design, determining navigation parameters in WGS-84 or PZ-90 (with a confidence level of 0.95).

The basic module contains the necessary hardware and a set of interfaces providing implementation of basic functions and technological modes.

The extension module has the additional functionality related to interaction with advanced intelligent systems installed on the vehicle's board.

The BIAS software has a multi-level modular hierarchical structure. Individual modules operate asynchronously with respect to each other, which allow realizing multi-tasking mode of operation in real time. The file structure used provides maximum isolation of the levels code and modules.

Research methods for the transition of vehicles from "automated" driving mode to "manual"

Until now, research methods for the transition of vehicles from the automated driving mode to the "manual" one stand at the beginning of development and are at the stages of "search" and research projects [8, 9, etc.]. The overwhelming majority of research projects use the methods for monitoring and evaluating the driver's visual analyzer dynamics applying appropriate experimental equipment in interrelations with the regaining of driver's control over vehicle's trajectory movement, situation conditions and performance of algorithms (or their fragments) in different variations and in real time with the vehicle automation levels from the first to the fourth according to the SAE's levels of driving automation.

It was found that depending on the specific traffic conditions, the tasks the driver was engaged in before the circumstances for resumption of the driving control arose (e.g. passive observation of the traffic situation, various telemetric services use, information on DVD media viewing), the driver will need some tens of seconds to regain control over the vehicle. But it indicates that "further research should identify how to inform the drivers in the best possible way of their commitment to resume control over the automated vehicle".

One of the common options can be a request from the driver for his readiness to regain control over the vehicle driving. The driver's subjective response is significant, as he/she clearly understands that the positive response gives him/her the entire measure of responsibility, including the responsibility for the road safety. Table 1 shows that ADAS of DMS reliability of the interpretation of the driver's functional state (driver's state control system) is classified according to the technology used in the particular DMS, and the probability of a dangerous failure lies in the range p– 02 – 0.001, using the methods of monitoring the driver's visual analyzer and p – 0.0001, using methods of monitoring the analysis of EDA dynamics with automatic interpretation of the driver's functional state (the ability to perform specified algorithms of activity) of "deep relaxation" type.

The methodology of monitoring of the functional state (the ability to perform the specified operational algorithms in real time, the readiness for emergency action) of "human-machine" transport systems operators, widely approved by locomotive drivers and car drivers [10-13, etc.] and based on automatic analysis of the parameters of the operator's EDA, is widely known. The systems for maintaining the performance of vehicle drivers have passed qualification tests [14] and continue developing [3-6].

The on-board module for determining the driver's hazardous states (deep relaxation, loss of consciousness or death of the driver) is used in the BIAS where automated interpretation of the driver's relaxation (not compatible with the required readiness to perform the algorithms of driving activity) is based on an automated analysis of the EDA parameters, which provides efficiency and the highest reliability of the proposed approach to the development and investigation of a new method(s) for estimating the potential for the transition (resumption) of driver's control over the vehicle by on-board systems in automatic mode.

Conclusion

Implementation in the BIAS of the functions of the driver's functional state monitoring, high-precision determination of the current vehicle navigational coordinates and analysis of heterogeneous current information (dynamic characteristics of the vehicle, the state of the main units and assemblies, driver state, navigation coordinate and its trend) in real time, where original algorithms and software are used, the modularity of the BIAS construction and the possibility of increasing the functionality through integration with other on-board systems has a development perspective and is aimed at improving the efficiency of "human-machine" motor vehicle systems.

The BIAS, being essentially an on-board integrating complex, can also be used as an experimental system to justify and develop the new methods for monitoring the ability of resumption of control over the vehicle by the driver, when the driver is potentially able to perform the required algorithms of activity (to regain control over the vehicle) while transition of control from automated driving to the "manual" one.

References

[1] SAE J3016 2014 Taxonomy and definitions for terms related to on-road motor vehicle automated driving systems (Warren dale, Society of Automotive Engineers)

[2] Valentin K. 2017 The future of the automotive industry: a vehicle defined by software Proc. Int. Automob. Sci. Forum "Intelligent Transport Systems" (Moscow)

[3] Savchenko V.V. 2013 Development of methodology for monitoring the functional states of operators of human-machine transport systems Mechatronics, Automation, Control 6 27–32.

[4] Bonch-Bruevich V.V. 2015 Remote control of driver's vigilance on route Automation in the industry 233–5

[5] Dementienko V.V. 2015 Driver vigilance remote monitoring system Science Journal of Transportation 6 110–14

[6] Savchenko V.V. and Poddubko S.N. 2017 Integration of passive assistance systems to the driver with on-board vehicle systems. "Forum Intelligent Transport Systems" (Moscow)

[7] NT1065/66_USB3 multi-channel GNSS RF front-end board. [Internet] 2017 Oct Available from: http://ntlab.com/section/sec:v:44976.htm

[8] Merat N. 2012 Highly automated driving, secondary task performance and driver state Human Factors 54 762–71

[9] Merat N., Jamson H., Lai F. and Carsten 2014 Transition to manual: Driver behavior when resuming control from a highly automated vehicle. Transportation Research Part F: Traffic Psychology and Behavior 27 274–82

[10] Savchenko V. V. 2005 Methods and means of increasing the efficiency of the functioning of "human-machine" transport systems operators News of the NAS of Belarus. Series of physical and technical sciences 2 9–37

[11] Dementienko V.V. 2008 Biomathematical model of human operator falling asleep. Physiology of man 34 (5) 63–72

[12] Dementienko V.V. 2010 Physical principles of construction of systems for safe monitoring of human operator status. Extended Abstract of D. Sc. Thesis (Moscow) p 41

IOP Conf. Series: Materials Science and Engineering **386** (2018) 012007 doi:10.1088/1757-899X/386/1/012007

[13] Dementienko V.V. and Dorokhov V.B. 2013 Evaluation of the effectiveness of the human operator's vigilance system with regard to the probabilistic nature of the occurrence of errors in falling asleep Journal of higher nervous activity 63 iss 124–32
[14] Savchenko V.V., Svistun M.S. and Sikorsky V.V. 2008 Driver maintenance system: test results and experimental research Automobile industry 132–4

IOP Conf. Series: Materials Science and Engineering **386** (2018) 012008 doi:10.1088/1757-899X/386/1/012008

Development of a detection road users system in vehicle A-pillar blind spots

P O Beresnev, A V Tumasov, D V Zeziulin, D M Porubov, L N Orlov

Nizhny Novgorod State Technical University n.a. R.E. Alekseev, Minin str., 24, Nizhniy Novgorod, 603950, Russian Federation

E-mail: pavel.beresnev@nntu.ru

Abstract. In connection with the increase in the number of cars in the world, there is an increase in the number of road accidents. The structural elements of the car do not allow the driver to assess fully the road situation. Such zones are called blind spots. Mostly the collision of cars happens in blind spots due to the lack of visibility. This is especially true for commercial vehicles. This article presents a system that allows you to expand the visibility of the car, due to the use of displays in the windshield pillars. A review of the existing systems was carried out. However, the use of only displays is not safe, for this, a pedestrian recognition system integrated in the windshield pillars based on neural networks can be used. The system allows detection, estimation of the distance to the object and giving out a warning signal about a possible collision. It consists of an on-board computer module, a video camera and an display. This system allows you to reduce the number of accidents near a pedestrian crossing or with a sudden pedestrian exit on the roadway. Interior elements have been designed and installed, allowing a good fitment of the displays into the windshield pillars. Exterior elements have been designed and installed in such way that allowed to keep the visibility of the car. The external side-view mirrors were replaced with video cameras. Experimental studies of the system were carried out. The pedestrian recognition system has demonstrated its performance by the results of the tests, however, it is necessary to debug the coefficients taking into account the approach time to the object.

1. Introduction

One of the important components for the creation of an efficient competitive transport industry is the intellectualization of transport systems. The development of intelligent transport systems is a relatively new direction in science and technology and represents an effective tool for solving transport problems, being a source of creating new branches in the industry.

The systematization of existing solutions in the field of developing of intelligent transport technologies, conducting their analysis for identifying the perspective directions, and developing a unified approach for the implementation of projects in this area is very topical. For many years, research related to the development of driver assistance systems (ADAS) has been carried out to improve road safety. The main algorithm of ADAS-systems during the motion is the collection of data from sensors around the vehicle, further analysis of information, detection of a critical situation on the road and conducting measures to avoid or mitigate the consequences. To perform these tasks, the car is equipped with various sensors (video sensors, laser sensors, ultrasonic sensors), audio and visual alerts, as well as an information output device (displays).

IOP Conf. Series: Materials Science and Engineering **386** (2018) 012008 doi:10.1088/1757-899X/386/1/012008

These systems are able to distinguish between traffic situations in which a driver's inattention can lead to a traffic accident. They can simultaneously notify about each of them during the motion with unique alerts using sound or pictograms. As the basic systems of assistance to the driver it is possible to point out: the road lane keeping assistant, detection of obstacles on a road lane, detection of road signs, parking assistant, detection of pedestrians, system of night vision, adaptive cruise-control, the control system of blind spots, etc.

According to the statistics [1], one of the most common types of road accidents is the accident when entering the blind spot of the car. This is especially important for commercial vehicles, as the blind spots of the car increases with the growth of its dimensions.

Currently, there are a number of systems commercially available on Mercedes, BMW, Volvo, Tesla, Volkswagen, Toyota, Nissan, Ford, etc. The systems have unique trade names, for example Volvo - BLIS; Ford - BLISTM; Porsche - SWA (Spurwechselassistent); BMW - LCW (Lane Change Warning); Audi - Side Assist. However, the general algorithm of actions is similar. When entering the blind spot of the car, the system informs the driver of the presence of interference in this area by visualization on the side or center rearview mirror or dashboard cluster.

Such areas are located behind, from the side and ahead of the car. When a neighboring car hits the blind spot it becomes not visible due to the design features. Blind spots on the sides during motion are very dangerous. More than 70% of accidents on the roads happen due to this fault.

A common driver's mistake is changing lanes when in the blind spot there is the car, it leads to a lot of accidents.

Also the presence of areas in the front of the car should be noted. These areas are artificially created by windshield pillars, a hood or a high driver's seat. When making maneuvers or driving in a tight environment, blind spots are the cause of an accident.

Currently, most accidents with people in Russia are caused by the driver's fault, however, as statistics show, for the last quarter of 2017 [1], an accident due to the fault of a pedestrian also takes place.

The major accidents caused by pedestrians are related to the failure to comply with the rules of crossing the carriageway established by the law. Also improperly parked cars near pedestrian crossings significantly worsen the situation, which leads to a reduction in the viewing angle and improper assessment of the traffic situation by the driver.

In order to exclude these shortcomings, work is underway to reduce or partially eliminate the blind spots of the car. For example, the Jaguar Land Rover group introduced the 360 Virtual Urban Windscreen system (Figure 1), which is designed to improve the visibility of cars. The system allows you to project on the windshield pillars an image of a part of the road that is hidden by these elements of the car's construction [2].

Figure 1. «Transparent» pillars of the Jaguar [2]

Pillars becomes transparent automatically, for example, during lane changing or near a crossroads. Together with the screens in the pillars and cameras, there is a projection display, which highlights the participants of the traffic surrounding the car. In addition, models of the British automaker in the future

IOP Conf. Series: Materials Science and Engineering **386** (2018) 012008 doi:10.1088/1757-899X/386/1/012008

will be able to show, for example, fuel prices, streets with the accompanying information and congestion.

Another development is made by Audi, presented at the Frankfurt Motor Show. It is a concept car Nanuk quattro with the complete lack of rear-view mirrors [3]. To output information from video cameras, displays integrated into the windshield pillars are used (Figure 2a).

Designers and engineers of Renault when creating concept car, the Initiale Paris, proceeded simpler: from all three cameras the image is displayed on a single screen, replacing the inside rear-view mirror (Figure 2b) [4].

It is also necessary to highlight the development of the Volkswagen XL1. The developers propose to use the rear view camera instead of the side mirrors [5] (Figure 2c).

a b c

Figure 2. (a) Concept car Nanuk Quattro [3], (b) Concept car Initiale Paris [4], (c) Volkswagen [5]

The value of blind spots in cars depends entirely on the overall dimensions of the vehicle and the driver's location. This area is heavily endangered by cargo vehicles, (van, truck, bus and etc.).

So the company Iveco at the motor show of commercial vehicles in Hanover introduced the concept of the hybrid van Vision [6] (figure 3a), with the displays installed in the windshield pillars.

Orlaco [7] proposes the introduction of MirrorEye's video surveillance system on trucks instead of rear-view mirrors (Figure 3b).

The innovative Orlaco system meets all the requirements and standards of EMC and is produced according to the guidelines for the automotive industry - ISO / TS 16949. At the request of car owners, MirrorEye's cameras and monitors will be available for all types of trucks.

a b

Figure 3. (a) Iveco Vision, (b) Orlaco truck

Based on the presented review of information and works [8-12], it was revealed that the use of video cameras instead of side-view mirrors is a promising direction. However, these cameras do not allow you to look ahead of the car, which in turn could reduce the risk of a dangerous situation associated with a pedestrian collision in the «blind» areas in front of the vehicle.

Based on these findings, the authors proposed a concept system that combined the functions of detecting objects in the blind spots in front of a light commercial vehicle and exterior side mirrors (Figure 4).

IOP Conf. Series: Materials Science and Engineering **386** (2018) 012008 doi:10.1088/1757-899X/386/1/012008

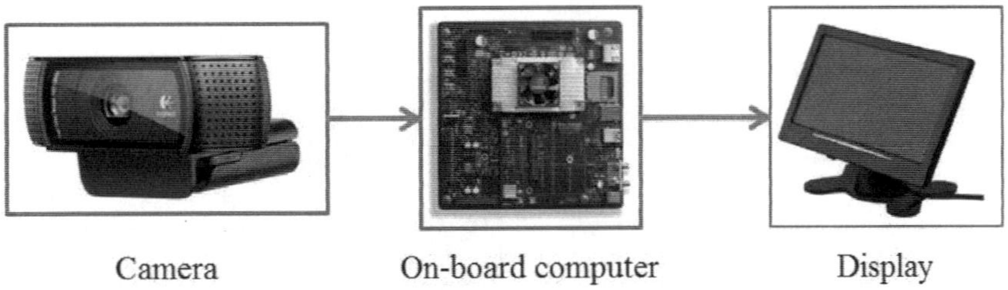

Figure 4. System interaction diagram

Initial studies were carried out in the framework of the work described in [13]. Based on the results of these works, it was proposed to integrate the displays into the windshield pillars of the Gazelle NEXT. This solution will ergonomically fit the displays into the interior of the car, as well as maintain visibility to the driver.

2. Development of the system

As can be seen from Figure 4, the system consists of a video camera that collects a video stream data for further processing; a computing module based on the platform for the development of Nvidia Jetson TX2. With the capabilities of the Nvidia Jetson TX2 [14], information from the system can be sent over the CAN bus to interact with the other (executive) devices of the vehicle. Also, the performance of this platform allows processing data for other ADAS-functions, which will be installed additionally.

The Jetson TX2 module is equipped with an NVIDIA graphics processor with Pascal architecture, a CPU Complex ARMv8 module with six cores. Module Jetson TX2 has a relatively small overall dimensions (50x87 mm), its weight is 85 g.

Logitech camera is used to solve problems of object recognition in a blind zone. It should be noted that the developed system can be used with various video cameras.

On-board computer NVIDIA Jetson TX2 is placed in a specially manufactured case, which is shown in Figure 5. This case allows to reduce the risk of damage from external influences (falls, impacts, etc.). Also in the case there is installed the air cooling system of the computational module to prevent overheating of the system.

Figure 5. Assembling the on-board computer module

The computational module collects information from cameras installed instead of external side-view mirrors. Its location is under the seat of the car. The module connects to the Can-bus of the car to collect data on the speed of the vehicle.

For the installation of displays, computer models of plastic car interior elements were developed (Figure 6).

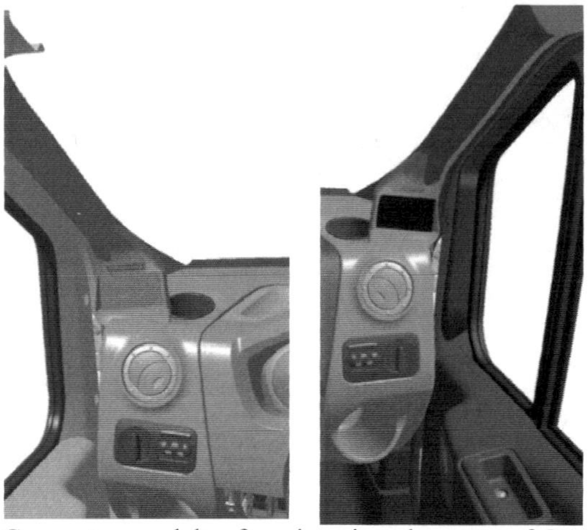

Figure 6. Computer models of car interior elements of GAZelle NEXT

At the next stage, the elements were made according to computer models. After manufacturing, the elements were installed in the car's interior (Figure 7).

Figure 7. Interior elements of the GAZelle NEXT

The installation of these elements allowed to fit the displays in the interior of the GAZelle NEXT. Also, the use of the elements allowed to maintain the visibility of the car. Displays were installed in terms of the convenience of the driver's perception of information, thereby increasing the ergonomics of the car's interior.

Also based on the review, it was suggested to use the system of video cameras instead of external side mirrors. The brackets for fixing the cameras outside the car has been made (Figure 8).

Figure 8. Brackets for attaching video cameras

Using cameras instead of external side mirrors gives a larger viewing angle, and the right angle allows the driver to correctly assess the traffic situation.

A number of experimental studies were carried out to evaluate the system's operability.

3. System Tests

The tests were carried out on the GAZelle Next vehicle with an installed pedestrian collision warning system. The system's algorithm consists in the following steps: The system determines the speed of motion and recognizes the object on the route with the help of a video camera, then calculates the distance to it.

To assess the performance of the system, a series of races was conducted, simulating the sudden exit of a pedestrian to the carriageway. The tests were carried out on both the right and left sides of the car, while the pedestrian was in a blind spot of the windshield pillars.

It should be noted that the braking was carried out after the start of the warning signal. A fragment of successful braking in front of a dummy simulating a pedestrian is shown in Figure 9.

Figure 9. Successful braking in front of a dummy simulating a pedestrian

The criterion for the correct operation of the system was the timely delivery of a warning signal to the driver about a possible collision with a pedestrian. The calculated accuracy of the determination of the critical situation, which characterizes how many of the positive triggering turned out to be really correct, was 0.91. The completeness, showing the number of correct responses from the total number of tests conducted, was 0.9. The indicator of the quality of the system was 89%.

Also test races were conducted without a warning system about a possible collision with pedestrians. The test conditions were preserved. Due to the impossibility of detecting a pedestrian by a driver in a blind spot, a collision occurred. This type of accident is mainly encountered when driving near a pedestrian crossing [1]. Fragments of tests carried out without a collision warning system with pedestrians are shown in Figure 10.

Figure 10. Work without a warning system about a possible collision with pedestrians

It should be noted that the main restriction of the use of cameras instead of external side-view mirrors is only the viewing angles of the cameras.

4. CONCLUSIONS

The system of visual information of the driver integrated into the windscreen pillars is presented, with the possibility of warning about a collision with pedestrians on the route. The use of video cameras instead of external side mirrors is considered.

Interior elements of the GAZelle NEXT vehicle were developed and installed. These elements have allowed to maintain visibility and increase the ergonomics of the car interior.

Experimental studies of the developed system have been carried out. The pedestrian recognition system has demonstrated its performance by the results of the tests, however, it is necessary to debug the coefficients taking into account the approach time to the object. In addition, the tests were carried out on a road covered with snow. This led to the fact that the braking distance, in comparison, with dry asphalt exceeds more than twice. This phenomenon requires detailed study and taking measures to minimize the influence of this factor on the correct operation of the system.

Acknowledgments

This research done with the financial support from Ministry of Education and Science of the Russian Federation (the unique project identifier is RFMEFI57717X0268).

The experimental research conducted with the use of measurement equipment of the NNSTU Centre of collective using «Transport Systems».

Reference

[1] Autostat. Analytic agency. Retrieved May 30, 2018, from: https://www.autostat.ru/
[2] Jaguar Land Rover. Retrieved May 30, 2018, from: https://www.jaguarlandrover.com/
[3] The world of Audi. Retrieved May 30, (2018), from: http://www.audi.com/en.html
[4] Renault FRANCE Constructeur automobile. Retrieved May 30, 2018, from: https://www.renault.fr/
[5] Volkswagen World. Retrieved May 30, 2018, from: http://www.volkswagen.co.uk/about-us/concept-cars/xl1
[6] IVECO TRUCK. Retrieved May 30, 2018, from: https://www.iveco.com/russia/company/pages/z-truck.aspx
[7] Orlaco Global. Retrieved May 30, 2018, from: https://www.orlaco.com/
[8] Fitch, G. M., Blanco, M., Camden, M. C., & Hanowski, R. J. (2011). Field demonstration of a camera/video imaging system for heavy vehicles. *SAE International Journal of Commercial Ve*hicles, 4(1), 171 - 184. doi:10.4271/2011-01-2245
[9] Fitch, G. M., Schaudt, W. A., Wierwille, W. W., Blanco, M., & Hanowski, R. J. (2011). Human factors and systems engineering of a camera/video imaging system. *In Proceedings of the 18th World Congress on Intelligent Transportation Systems*. Washington, DC: ITS America.

[10] Schaudt, W. A., Wierwille, W. W., & Hanowski, R. J (2008). Enhanced camera/video imaging systems (E-C/VISs) for heavy vehicles. *SAE International Journal of Commercial Vehicles* 1(1), 293-304.
[11] Camden, M. C., Fitch, G. M., Blanco, M., & Hanowski, R. J. (2011). Commercial driver acceptance of heavy vehicle camera/video imaging systems. *In Proceedings of the Human Factors and Ergonomics Society 55th Annual Meeting. Santa Monica,* CA: Human Factors and Ergonomics Society.
[12] Schaudt, W. A., Bowman, D. S., Hanowski, R. J., Olson, R. L., Marinik, A. R., Soccolich, S., Joslin, S., Toole, L., & Rice, J. C. (2014, February). FMCSA's Advanced System Testing utilizing a Data Acquisition System on the Highways (FAST DASH) program: Safety technology evaluation #1 blindspot warning: Final report (Report No. FMCSA-RRT-13-008).
[13] Development of driver's assistant system of additional visual information of blind areas for Gazelle Next. 2017. V Makarov, O Korelin, D Koblyakov, S Kostin and A Komandirov, IOP Conference Series: Materials Science and Engineering, Volume 315, conference 1
[14] The Official NVIDIA Blog. Retrieved May 30, 2018, from: https://blogs.nvidia.com/

IOP Conf. Series: Materials Science and Engineering **386** (2018) 012009 doi:10.1088/1757-899X/386/1/012009

Development of a mobile robot group for coastal monitoring

D Yu Tyugin, D V Zeziulin, A A Kurkin, V V Belyakov, V S Makarov

Nizhny Novgorod State Technical University n.a. R.E. Alekseev, Minin str., 24, Nizhniy Novgorod, 603950, Russian Federation

E-mail: pavel.beresnev@nntu.ru

Abstract. In Russian Federation rich oil and half of all gas reserves are distributed along the sea shelves. Severe weather conditions, frequent storms and intense sea waves, complicate exploring and mining of new shelf deposits. Therefore nowadays stationary automated systems are used for monitoring the environment in the shelf zone of the Sea of Okhotsk near Sakhalin Island. But they provide data only for specific areas of operation. Such restrictions can be removed by using groups of remotely controlled mobile robots. Moreover, mobile vehicles should be able to operate under any weather conditions. This paper summarizes the research results of the development and creation of efficient chassis designs for coastal areas of Sakhalin Island and adjacent waters. The technique for choosing vehicle design parameters and assessing the efficiency of the system as a whole is given. Data on physical and mechanical characteristics of soil and gravel bases in the coastal zone were used for estimating vehicles' cross-country ability. To create a multi-agent coastal monitoring system we intend to use several ground, underwater and overwater robots able to exchange information in real time. We give an outline of a mobile robot group operation, the group being composed of ground, underwater and overwater robots. The use of mobile robots for performing the tasks mentioned above has the following advantages: expansion of the working area achieved by allocating the robots; highly probable accomplishment of the task due to the possibility of redistribution of targets among group members; the ability to take measurements at several locations simultaneously.

1. Introduction

Constantly growing state demands for fuel, energy, various mineral raw materials are hastening the using of ocean energy resources for the state economy development. Shelf zone plays an important role in maintaining world oil and gas production. Over the past ten years, more than two-thirds of the hydrocarbon reserves were discovered in shelf zones. The Arctic shelf is of particular importance, the total volume of its oil and gas reserves is about 413 billion barrels of oil equivalent (about 22% of the total unexplored stocks of traditional hydrocarbons in the world). All Arctic states enacted laws confirming strategic importance of the Arctic primarily in terms of hydrocarbon reserves. The Russian Federation is one of the leaders in the development and implementation of the shelf zone mineral extraction projects. The first project running on the Russian Arctic shelf in the Pechora Sea is the Prirazlomnoye petroleum deposit development. Its oil reserves are assessed at 72 million tons. The offshore ice-resistant oil production platform Prirazlomnaya has full capacity of up to 6.5 million tons per year. At present one of the most important areas of deposit active development is the Sakhalin Island shelf zone. Piltun-Astokhskoye and Lunskoye fields on the north-eastern shelf of the island are being developed by the Sakhalin Energy Investment Company Ltd [2], one of the largest oil and gas companies. To reduce production, financial and environmental risks prompt and accurate assessment of a working zone should be made at each design and operation stage. With the advancing of the projects on Russian shelf zone energy resources development ,problem of coastal zone monitoring is becoming more and more topical.

To collect information on the coastal zone, aircraft and space-based radar stations are widely used. However, common aerospace methods do not provide the required accurate and rapid information on a coastal zone concerning wave dynamics. Without such information, safe operation of shore facilities is impossible.

The main task is to monitor the wave situation in areas where oil and gas facilities are being operated using ground-based radar systems to refine the data obtained via space means. Thus, at the present development stage of radar methods for wave climate estimation it is important to observe the wave activity in the surf zone constantly, in any weather conditions.

Another way is to monitor the coastal zone using aircrafts. However, this method has a number of limitations on load capacity, safety and range.

As a result, to monitor the wave climate in the coastal zone ground vehicles moving along the bottom of the water area for are widely used [3, 4]. Currently, to detect distance to the object in coastal zone survey remote sensing methods using sensors based on the radio [5] or laser waveband [6, 7] scanning devices are considered to be prospective ones.

2. Movement conditions

Movements within coastal zones are restricted due to convertible humidity of the supporting surfaces as well as to the obstacles along the road (stones, sediments, etc.).

Analysis of movement conditions allowed to elaborate a classification of movement surfaces in coastal zones as roadways for transport and technological machines.

Sand-gravel and sandy areas, as well as snow mantle in winter prevail as supporting surfaces in coastal zones .

Figure 1 shows structure and classification characters of coastal zones.

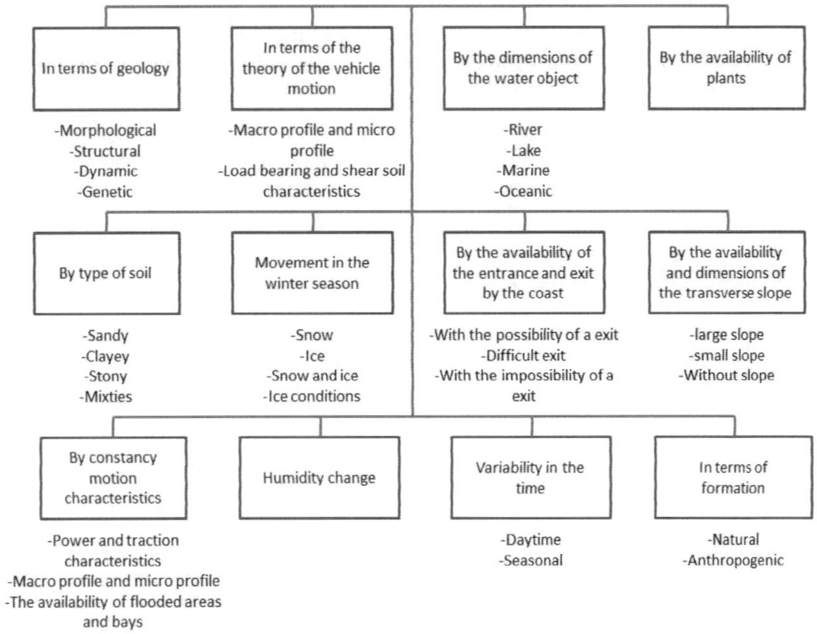

Figure 1. Structure and classification features of coastal zones.

In 2016 and 2017 the authors carried out research of the coastal zone of the Sea of Okhotsk, Mordvinov Bay [8, 9].

The most convenient mathematical model for sand-gravel supporting surfaces is given in the works of Y.S.Ageikin and N.S.Volsky [10]. Normal and shear stresses calculation is made according to the dependencies given below:

$$q_\beta = \left[\frac{(H_G - z)}{b(1+1{,}75\varphi_0)(k_{\beta_1}\cdot b\cdot \rho \cdot X_1 + k_{\beta_2}\cdot c_0 \cdot X_2 + k_{\beta_3}\cdot \rho \cdot X_3 \cdot z)\cdot \cos\beta} + \frac{a\cdot b}{E\cdot z} arctg \frac{(H_G - z)}{a\cdot b\cdot \cos\beta} \right]^{-1} \quad (1)$$

$$\tau = \left[q_\beta \cdot tg\varphi_0 + c_0 \left(1 - \frac{S_{t_0}}{S_t}\right) \right] \left[1 - \exp\left(- \frac{S_{t_0}}{k_\tau} \right) \right] \quad (2)$$

где $k_{\beta_1} = \dfrac{\pi - 4\cdot\beta\cdot tg\varphi_0}{\pi + 4\cdot\beta\cdot tg\varphi_0}$, $k_{\beta_2} = k_{\beta_3} = \dfrac{3\cdot\pi - 2\cdot\beta}{3\cdot\pi + 2\cdot\beta}$, $X_1 = \dfrac{1 - X^4}{X^6}$, $X_2 = \dfrac{1{,}3[X^2 + 1]}{X^4}$, $X_3 = \dfrac{1}{X^6}$

где $X = tg(\pi/4 - 0{,}5\varphi_0)$

Dependences for calculating the generalized functions of resistance and friction are determined on the basis of the values of the stresses arising in the elemental area with subsequent integration over the contact area.

$$\Phi_f = b M_a^{-1} \int_0^{h_G} p dh, \quad (3)$$

$$\Phi_\varphi = b M_a^{-1} \int_A \tau dA. \quad (4)$$

Figure 2 shows calculation schemes for wheeled and track movers.

Ratio analysis for the generalized functions of resistance and friction allows assessing chassis's ability to move in coastal zone over sandy and sand-gravel support surfaces and defining such values of design parameters when chassis will remain movable. In accordance to this, we developed a methodology, its block diagram is shown in Figure 3.

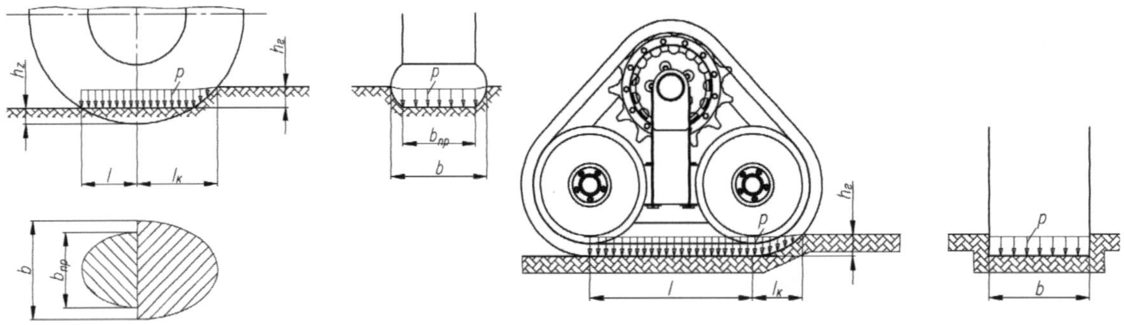

Figure 2. Design diagram of wheeled and track movers' interaction with soil.

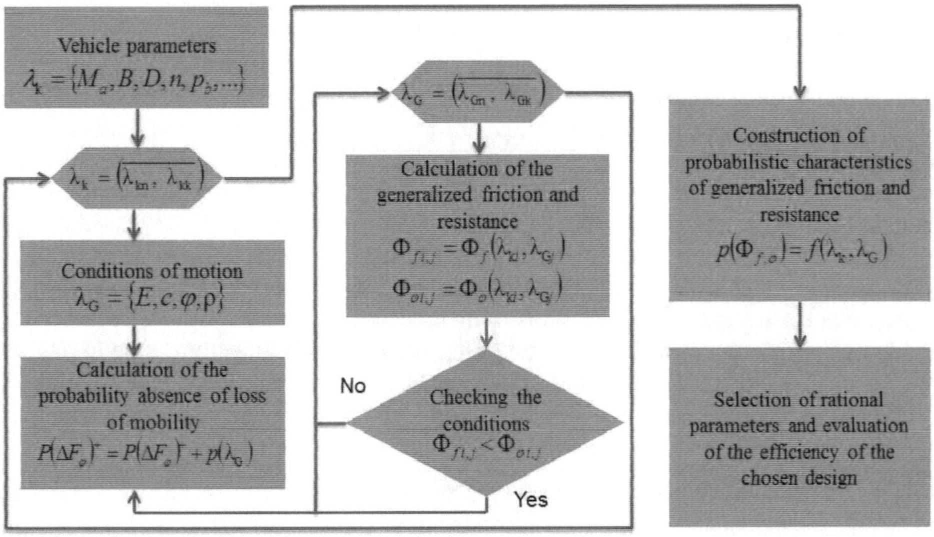

Figure 3. Block diagram of the methodology for evaluating the structural parameters of the chassis and efficiency criterion defining.

The suggested methodology uses loop with a counter (searching), base parameters of the chassis $\lambda_k = \{M_a, B, D, n, p_b, ...\}$ change their value from the specified initial value λ_{kn} to the final value λ_{kk} with some step $\Delta\lambda_k$ the loop body is executed once for each parameter. The initial values of the parameters λ_k are preset on the basis of engineering experience and the design requirements provided by the requirements specification. In this paper we analyze the change in the parameter values by 20% from the average:

As data on the characteristics of the supporting surface $\lambda_G = \{E, c, \varphi, \rho\}$ the values, as well as their statistical characteristics, are set, which are set using a cycle with a counter (enumeration). The basic parameters of supporting surface λ_G change their value from the specified initial value λ_{Gn} to the final value λ_{Gk} with a certain step $\Delta\lambda_G$, the loop body is executed once for each parameter.Meanwhile, each value λ_G corresponds to its own probability density $p(\lambda_G)$.

For each parameter ratio λ_G and λ_k, the values of the generalized resistance Φ_f and friction Φ_φ functions are calculated.If $\Phi_f < \Phi_\varphi$, we calculate the coefficient that takes into account vehicle's mobility being preserved under practicability condition with λ_G and λ_k given. By varying the parameters of the ground λ_G and obtaining the values of Φ_f and Φ_φ, we represent dependencies which take into account the probabilistic characteristics in the form $p(\Phi_f) = f(\lambda_G, \lambda_k)$ and $p(\Phi_\varphi) = f(\lambda_G, \lambda_k)$ for given values. λ_k [11].

To estimate and choose rational parameters of the movers for the developed chassis when operating in snow-covered terrain , we used mathematical models described in [12, 13].

The method serves for the development of mobile ground complexes for monitoring coastal zones, which are necessary for ensuring activity in coastal areas and adjacent water areas. The block diagram is shown in Figure 4. Using this method one can recommend a certain chassis design (choose mass and dimensions parameters, dimensions of the mover), evaluate the effectiveness of the selected algorithms for controlling the movers' power distribution flow, determine the fields of the chassis's efficient use for wheel and track movers in warm seasons and in winter [14, 15].

The sequence of actions is as follows. The requirements for the place and time of monitoring and research equipment specification are developed in terms of safety requirements for this or that activity. Knowing place and time of monitoring, one is able to qualify the surface on which the research complex will operate. It can be either sand-gravel or snow-covered terrain. Additional sample probes of the coast can taken as needed. Equipment specification defines the dimensions of the chassis and the weight of cargo being transported. It will determine the basic parameters for the chassis on the basis of regression equations for the vehicles ratio. Having chosen the initial data, we analyze the constructions and choose design parameters optimal for the set operation conditions. At first, we evaluate the design from the point of view of its cross-country ability and choose the most reasonable alternative. Then, we make estimates and select algorithms for controlling the movers' power distribution flow. Using the obtained data, we can estimate autonomous travelling time for the vehicle with lowest fuel consumption. The mobility of the chassis is evaluated reasoning from its potential travelling time during the winter with cross-country ability preserved. Thus, the sequence of actions given helps to suggest chassis designs optimal for mobile monitoring sets operating in coastal zones.

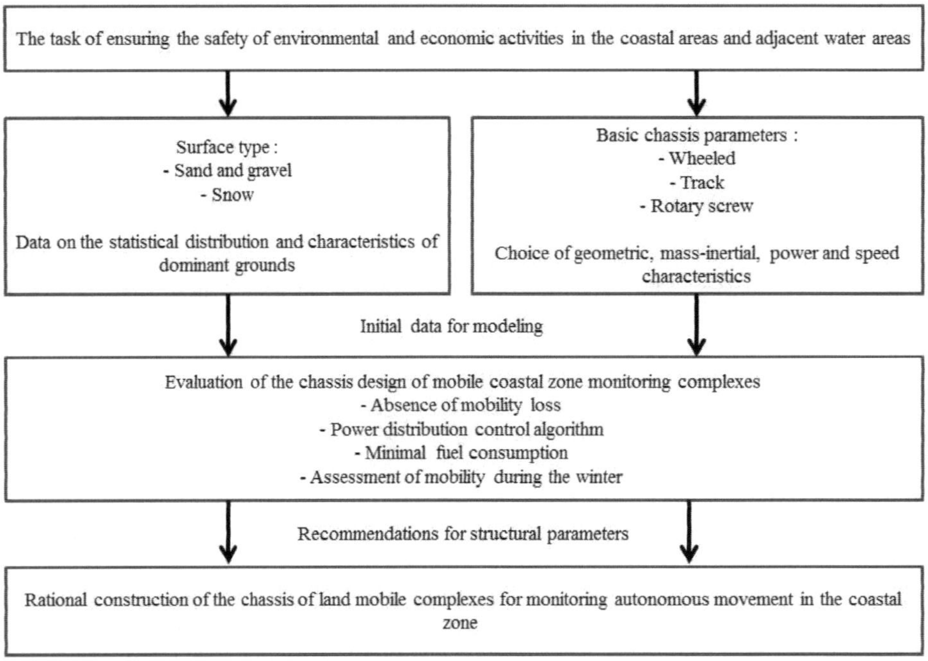

Figure 4. Block diagram of the method for designing the chassis of mobile monitoring sets.

3. Mobile robot group

In accordance with the proposed method, various technological vehicles and mobile robotic complexes were developed, including an autonomous mobile robotic complex [16, 17, 18, 19] for coastal zone monitoring. Experimental researches of this complex carried out on the coast of Sakhalin Island in May and June 2016 made it possible to evaluate the efficiency of the worked out technical solutions and algorithms for the mobile monitoring systems operation in coastal zones.

However, using single robot for research work is inefficient. For this purpose, we suggest creating a group of mobile robots capable of monitoring the wave conditions in the shelf zone and transporting measuring equipment for complex research. Methods based on using of a group of remotely controlled mobile robotic means are free of many faults mentioned above, and proved their efficiency in the course of experimental researches [20, 21, 22].

IOP Conf. Series: Materials Science and Engineering **386** (2018) 012009 doi:10.1088/1757-899X/386/1/012009

Within this project we suggest an approach to coastal wave climate monitoring ,based on employing a telemetrically connected group of ground, above-water and underwater robots.

This approach allows us to expand the research area significantly.

The scheme of the group interaction and the possible scheme for carrying out the experiment is shown in the Figure 5.

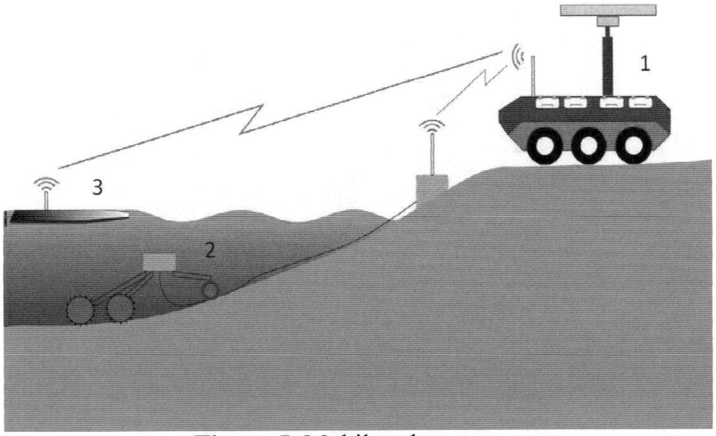

Figure 5. Mobile robot group.

This scheme consists of ground (1), underwater (2) and above-water (3) robots. The ground and above-water complexes exchange information on wave strength and the investigated area bathymetry detailed data. The underwater robot uses a cable to transmit data on wave strength from a string sensor in the researched area,the data is received by a stationary base installed on the shore. Further, the information is transmitted via wireless communication to the ground robot computer for processing. Now, we'll examine each type of robot.

3.1. Ground robot

In the interaction scheme, the ground segment is represented by an autonomous mobile robotics facility developed by research team from NNSTU n.a. R.E. Alekseev [23, 18]. The complex,which shown in Figure 6, can have different types of movers installed, add-ons for actuating devices of control drives and hardware.

The hardware consists of two LMS511Pro LIDARs, a GNSS antenna, a weather station, a wi-fi antenna, a video camera and all-round surveillance radar MRS-1000.

The robot collects information on wave activity in the investigated zone. Software unit collects information and outputs it graphically. During the tests the data is all held on the on-board computer.

Figure 6. Ground robot.

3.2. Underwater robot

The underwater segment in the scheme is represented by an amphibious modular vehicle (AMV) (Figure 7) [14]. AMV is a three-support system consisting of several basic components: support platform, sealed container for measuring equipment, adjustable arms, step motors, driving wheels and a supporting wheel.

Figure 7. General view of AMV [14].

The complex collects data using a video survey system and a hydrostatic wave recorder with a string sensor mounted to the stationary casing support. The operation of the string sensor is the work of the capacitor coatings between the insulator, water and copper wire being conductors[14].

AMV has adjustible gravity center for stability increasing when driving on steep slopes of the seabed or suffering significant hydrodynamic exposure. To adjust it vertically one changes the length of each of the support arms in the range of 770-920 mm, or changes the angle between support arms of the driving wheels in the range of 40°- 120° if horizontal adjustment is needed. When the position of the mass center in the two above-mentioned planes changes, the motion axes of the driving wheels diverge. This leads to disalingment of propelling forces, drag forces increase and loss of complex controllability. To avoid this problem we use an adjustment unit that allows changing the direction of the motion axes of the driving wheels and setting them parallel to each other with accuracy of ± 1 degree.

The AMV electronics consists of control and power units. A single-board computer makes the core of the control part,all other sensors are connected to it. These are tilt sensor based on MPU-6050 gyroscope, string sensor for recording wave height and videocamera. They are controlled remotely from the shore via Ethernet cable or wi-fi if the cable is connected to a buoy floating in the water. The power part consists of two stepper motors with reducers (1:10) and stepper motor drivers. Lithium battery 48V 10Ah is a power supply providing 4 hours of continuous operation which satisfies several submersion experiments.

3.3. Above-water robot

The autonomous remote-controlled research boat is designed to carry out detailed bathymetric measurement of the observable water area. It allows us to create a depth map promptly using a multibeam echosounder.

The development of an autonomous remote-controlled research boat was carried out by the engineers of Special Design Office for Automation of Marine Research of the Far Eastern Branch of the Russian Academy of Sciences (SRBA MR FEB RAS) who have wide experience in creating research complexes and conducting oceanological research, studing hydroacoustic, hydrophysical and hydrodynamic processes, the atmosphere and ocean interaction, marine hazards and marine safety.

For the positioning of the boat, the Trimble SPS461 GNSS receiver is used, the IMU-108-30 SMC is used to determine the boat's tilt parameters. Bathymetric data is collected by a multibeam echosounder. Protected computer with a separate power supply 12V 15Ah processes the data and controls the boat. Power plant includes two DC motors, BTS7960 drivers and an arduino nano controller. The boat is controlled remotely, maneuvering is performed through the right and left motors manipulating. Extended-range Wi-Fi module based on the MikroTik 2SHPN point of presence provides communication with the boat .

4. Experiment

We undertook three expeditions to various parts of the Sea of Okhotsk coastal zone at Svobodny Cape of Sakhalin Island and carried out AMV and AMRC tests in July 2017.

The coastal zone relief is characterized by shallow water depth of 2-4 meters over several hundred meters far from the shore. The AMV was plunged into water to the depth of about 2 meters using remote control at a distance of about 50 meters from the shore. At the same time, we collected data on wave activity using a ground robot placed near the AMV (Figure 8). The data from the string sensor was transmitted to a base station installed on the shore, and then to AMRC, that made equipment calibration possible, and we got information on wave strength in the zone with the help of radar.

Thus, both robotic devices were combined into united local network. Then, the data on wave height was collected and displayed in real time for 60 minutes.

Figure 8. AMV's submersion into water.

On the basis of the experimental researches undertaken , the following conclusions were made.

The developed AMV can be used to verify data received from a radar station installed on a large-scale research chassis when using a group of mobile ground and above-water robots for detailed study of wave dynamics.

Primordially, a string wave sensor was used, but this sensor has a number of shortcomings.It can be used only in a slightly rough sea and there is a risk that marine growth affects the delicate wire. In this connection , we plan to refine the AMTS design and use bottom pressure sensors, which do not have these drawbacks.

We made coastal supporting surfaces classification and elaborated an approach to the development of functional chassis designs for coastal and adjacent water areas.

A technique for selecting optimal vehicle parameters and basic vehicles requirements,depending on the type of terrain, have been developed.

We suggest brand new approach to sea waves rating with the help of a group of mobile ground, underwater and above-water robots. Unified approach will help expand the observable zone significantly and verify data received from robots.

For the summer 2018 we are planning an experiment involving all three types of robots integrated for better control into a local network with real-time data processing. Currently, software for shared control of a group of robots is being developed. We are planning to refine the system by increasing underwater robot mobility.Using the buoy with WiFi transmitter instead of underwater cable is possible. This solution requires underwater robot's connection to the buoy via cable, communication with other robots of the group will be established with the help of a Wi-Fi transmitter floating in the water. It will make the experiment less restricted since transmitter installation ashore and cabling won't be needed. It is also possible to install GNSS inside underwater robot, attaching the receiver antenna to the buoy, as a result this will improve the accuracy of robot's positioning.

Nomenclature

a	coefficient characterizing the attenuation of stresses in the soil	
b	tire contact width with soil	[m]
k_τ	coefficient of tangential elasticity of soil	
S_t	grousers step	[m]
S_{t_0}	grousers shear	[m]
β	angle between the load vector and the normal to the soil surface	[radian]
ρ	volume density of soil	
E	modulus of deformation	
φ_0	angle of internal friction	
c_0	internal friction in the soil	
H_G	thickness of soft layer	
z	immersion	
λ_k	basic chassis parameters	
λ_{kn}	initial value of basic parameters	
λ_{kk}	final value of basic parameters	
λ_G	basic soil parameters	
λ_{Gn}	initial value of basic soil parameters	
λ_{Gk}	final value of basic soil parameters	
$\Delta\lambda_G$	step values of basic soil parameters	
$\rho(\lambda_G)$	probability density	
Φ_f	generalized resistance function	
Φ_φ	generalized friction function	

Acknowledgements

This study was initiated in the framework of the state task program in the sphere of scientific activity of the Ministry of Education and Science of the Russian Federation (projects No. 5.4568.2017/6.7 and No. 2.1433.2017/4.6) and financially supported by this program and grant of the President of the Russian Federation No. NSh-2685.2018.5.

References

[1] Shelf Gazprom. Retrieved February 20, 2018, from: http://shelf.gazprom-neft.ru/business/prirazlomnoe_field/

[2] Official site of Sakhalin Energy. Retrieved February 20, 2018, from: http://www.sakhalinenergy.ru/ru/

[3] Modular Amphibious Research Crawler. Retrieved February 20, 2018, from: http://my.fit.edu/~swood/AUV_Crawler1.html
[4] The SPROV'er. Retrieved Febru-ary 20, 2018, from: http://my.fit.edu/~swood/SPROVER/The%20SPROV'er.htm
[5] Cheng H. and Chien H., 2017. Implementation of S-band marine radar for overwater wave measurement under precipitation, Remote Sensing of Environment, vol. 188, pp. 85-94.
[6] Park H.S., Sim J.S., Yoo J. and Lee D.Y., 2011. Breaking Wave Measurement Using Terrestrial LIDAR: Validation With Field Experiment on the Mallipo Beach, Journal of Coastal Research, 64, pp.1718-1721.
[7] Martin K., Bonneton P., Frappart F., Detandt G., Bonneton N. and Blenkinsopp C.E., 2017. High Frequency Field Measurements of an Undular Bore Using a 2D LiDAR Scanner, Remote Sensing, vol. 9, pp. 2-14, doi:10.3390/rs9050462.
[8] Mobile complexes for monitoring coastal zone: monograph / V.V. Belyakov, U.S. Vakhidov, D.V. Zeziulin, V.E. Kolotilin, A.A. Kurkin, V.S. Makarov, A.V. Tumasov, D.Yu. Tyugin; Nizhny Novgorod. state. tech. rniversity n.a. R.E. Alekseeva. - Nizhny Novgorod, 2017. - 326 p.
[9] Kurkin, A., Makarov, V., Zeziulin, D., Beresnev, P., Filatov, V., & Porubov, D., 2017. Study of coastal soil surfaces of sakhalin island. Paper presented at the 13th International MEDCOAST Congress on Coastal and Marine Sciences, Engineering, Management and Conservation, MEDCOAST 2017, , 2 775-785
[10] Ageikin Ya.S., 1972. All-terrain wheels and dual propellers. Theory and Design. Mashinostroenie. Moscow.
[11] Makarov, V., Filatov, V., Vahidov, U., Kurkin, A., and Belyakov, V., 2017. Study of trafficability conditions of typi-cal soils of coastal zones of sakhalin island (russian federation). Paper presented at the 19th International and 14th European-African Regional Conference of the ISTVS
[12] Porubov, D., Makarov, V., Zeziulin, D., Belyakov, V., and Anikin, A., 2017. Study of efficiency of using all-terrain vehicles during the winter period. Paper presented at the 19th International and 14th European-African Regional Con-ference of the ISTVS
[13] Anikin, A., Belyakov, V., Zeziulin, D., and Makarov, V., 2017. Calculation of traction capabilities of wheeled vehi-cles on low-pressure tires on snow. Paper presented at the 19th International and 14th European-African Regional Con-ference of the ISTVS
[14] Kurkin, A., Makarov, V., Zeziulin, D., Tyugin, D., Beresnev, P., Filatov, V., & Porubov, D., 2017. Research complex for surf zone analysis. Paper presented at the 13th International MEDCOAST Congress on Coastal and Marine Sciences, Engineering, Management and Conservation, MEDCOAST 2017, , 2 787-794.
[15] Zeziulin D., Beresnev P., Filatov V., Makarov V., Kurkin A., Belyakov V., 2016. Development of an unmanned ground vehicle for coastal monitoring, Proceedings of the 8th Americas Regional Conference of the ISTVS.
[16] Kurkin, A., Pelinovsky, E., Tyugin, D., Giniyatullin, A., Kurkina, O., Belyakov, V., Makarov, V., Zeziulin, D., Kuznetsov, K., 2015. Autonomous robotic system for coastal monitoring. 12th International Conference on the Mediterranean Coastal Environment, MEDCOAST 2015. 2., 933-943.
[17] Kurkin, A., Tyugin, D., Belyakov, V., Makarov, V., Zeziulin, D., Minaev, D., & Zaytsev, A., 2017. Multiagent net-work system for coastal monitoring. Paper presented at the 13th International MEDCOAST Congress on Coastal and Marine Sciences, Engineering, Management and Conservation, MEDCOAST 2017, , 2 795-804.
[18] Zeziulin, D., Makarov, V., Porubov, D., & Kurkin, A., 2017. Development of a ground mobile robot for motion in conditions of coastal zones. Paper presented at the 19th International and 14th European-African Regional Conference of the ISTVS. Paper presented at the 19th International and 14th European-African Regional Conference of the ISTVS.

[19] Kurkin A., Pelinovsky E., Tyugin D., Kurkina O., Belyakov V., Makarov V., Zezulin D., 2017. Unmanned ground vehicles for coastal monitoring, International Journal of Imaging and Robotics, V. 17. P. 64-75.
[20] Incoul A., Nuttens T., De Maeyer P., Seube N., Stal C., Touzé T. and De Wulf A., 2014. Mobile laser scanning of intertidal zones of beaches using an amphibious vehicle, INGEO 2014: 6th international conference on engineering surveying, Prague, Czech Republic, pp. 87-92.
[21] Wood S., 2006. Modular Amphibious Research Crawler, Sea Technology, vol. 47, no. 2, pp. 71-77.
[22] Wübbold F., Hentschel M., Vousdoukas M. and Wagner B., 2012. Application of an autonomous robot for the collection of nearshore topographic and hydrodynamic measurements, Coastal Engineering Proceedings, vol. 1, no. 3, pp. 2133-2143.
[23] Kurkin, A.A., Tyugin, D.Yu., Kuzin, V.D., Chernov, A.G., Makarov, V.S., Beresnev, P.O., Filatov, V.I. and Zeziulin, D.V., 2017. Autonomous Mobile Robotic System for Environment Monitoring in a Coastal Zone, Procedia Computer Science, 103. 459 – 465.

IOP Conf. Series: Materials Science and Engineering **386** (2018) 012010 doi:10.1088/1757-899X/386/1/012010

Development of a technique of gear ratios optimization with the aim of simultaneously increasing the vehicle dynamic, fuel efficiency and reducing emissions

S M Ogorodnov, P S Rogov, S I Maleev, Y P Kuznetsov, S A Plotnikov

Nizhny Novgorod State Technical University n.a. R.E. Alekseev, Minin str., 24, Nizhniy Novgorod, 603950, Russian Federation

E-mail:petr.rogov@nntu.ru

Abstract. The article deals with the problem of choosing of rational gear ratios with the aim of simultaneously increasing the vehicle dynamic, fuel efficiency and reducing emissions. A review of known approaches to the optimization of vehicle gear ratios is conducted. Russian Federation regulations requirements regarding vehicle dynamic are analyzed. The characteristics of an internal combustion engine, such as a mechanical and fuel map are obtained experimentally. The numbers of gears to be optimized are chosen reasonably. The methodology of the fuel usage calculating according to driving test in the standardized cycles was developed and demonstrated.

1. Introduction

Among the performance properties (PP) of vehicles, the dynamic properties and fuel efficiency are recognized as the most significant. It largely determines the effectiveness of vehicle and forms the consumer appeal and competitiveness. In recent years, the number of harmful substances (HS) in exhaust gases, emitted into the atmosphere during motion of vehicle equipped with an internal combustion engine (ICE), has acquired considerable importance. The values of emissions are standardized and listed in UNECE Regulations No. 83 [1] and No. 49 [2]. If these Rules are not met, vehicles are denied access to the sales market.

There are many possible ways simultaneously increasing of dynamic properties, fuel efficiency and at the same time reducing emissions vehicles. This article deals with optimization of vehicle gear ratios. This option is less expensive in comparison with engine upgrading or weight reduction, and also affects all three specified vehicle technical characteristics. The results of investigation can also be useful as basic data for algorithms of switching of automatic transmission gives additional value for work.

To determine the best approach to improving the vehicle performance indicators mentioned above, a review of existing research in this field was carried out. The works aimed at finding of optimum gear ratios, optimum algorithms of gear shift in automatic transmission reducing fuel or consumption of energy of vehicles were the center.

The aim of the article [3] was to search for optimal gear ratios to minimize fuel consumption and improve dynamic performance. Experimentally obtained detailed characteristics of the diesel engine were used in the investigation. Weight coefficients were used to combine the two optimization goals.

As a result, the Pareto curve was built to analyze the fuel efficiency and driving time. Since the subject of the study was a 30-ton quarry truck, the article considered only one cycle of its movement, in which there was no effect of stopping at traffic lights and traffic congestion, limiting the speed of traffic. In the article [4], specialists of General Motors conducted a study aimed at optimizing some parameters of automatic transmission, such as the number of gears, and the ratio of the top and first gears in order to reduce fuel consumption. An increasing the number of gears above eight was found impractical. Further increasing the number of gears does not lead to significant fuel economy. The minimum fuel consumption was achieved with number of gears of automatic transmission equals 10. In the article [5], optimization of transmission gear ratios was considered to minimize fuel consumption on a given route with specified macro profile parameters. The optimization was carried out using an algorithm based on solving specially obtained differential equations. To simplify the optimization algorithm the fuel consumption in the task was replaced by the energy consumption. The way of optimizing the gear ratios of a two-stage gearbox of a car, equipped with an electric motor as an engine, was shown in the presentation [6]. The optimization goals were the increase of dynamic indicators (acceleration time up to 100 km/h), as well as minimization of fuel consumption. For the second indicator, three standardized cycles were analyzed: NYCC, NEDC, HWFET. To find the optimal solution for two mutually opposite goals, energy consumption graphs as well as the acceleration time were plotted for each cycle, depending on the range of gear ratios of the two-stage reducer. The presented method is simple enough to implement and is usable for discussion with manufacturers. Particular interest from the point of view of the applied approaches to solving optimization problems is in the work [7], where the problem of minimizing the power consumption of a hybrid vehicle engine was solved due to the optimal energy management strategy. In the article [8] the problem of optimizing the gear ratios of the 6-speed gearbox was solved in order to reduce fuel consumption only. The Taguchi engineering method (TQEM) combining mathematical tools and tools for working with statistics was used in the experimental studies. In the work [9] simultaneous improvement of fuel efficiency and dynamic properties was achieved by developing several scenarios of optimal gear shifting in automatic transmission. The method of gear shift based on stochastic dynamic programming was found to be the most effective, which allows using it in real time mode. In the article [10] the problem of undesirable or superfluous switching of automatic transmissions, which is a consequence of that switching algorithms have no possibility to predict a road situation, was solved. To solve this problem, a gear shift prediction algorithm, with a route data from the navigation system and an additional data of the vehicle state, was used. The article [11] describes the activities that were carried out to increase the dynamics, reduce emissions of harmful substances and improve the fuel economy of the SUV. The first action to improve these indicators was the modernization of the ICE, namely the injection system, turbocharger and recirculation of exhaust gas. The second action was the selection of gear ratios for the purpose of satisfying predetermined indicators of dynamic properties, fuel efficiency and emissions of harmful substances. The NEDC and UDDS cycles were considered. It should be noted that only the first, fifth and main gears were considered in the optimization.

According to the conducted researches it is possible to make a conclusion that the topic of searching for the optimal gear ratios is quite well studied. The problems of the present study are closely connected with the problems of the work [11] with the exception that the goal of the work [11] was the satisfaction of given indicators, and not the search for the most effective gear ratios. To solve the problem of simultaneous reduction of fuel consumption and increase of the dynamics the approaches described in article [6] considering six gears instead of two and also adding another optimization goal for reducing the emissions can be found appropriate Taking into account the abovementioned, the distinguishing feature of this work is the extension of the optimization range of the gear ratios for six gears in conjunction with the emissions as an additional optimization objective.

To carry out the optimization of the transmission gear ratios, the following tasks must be solved:

- Preparation of input (initial) data, including the results of experimental studies.

IOP Conf. Series: Materials Science and Engineering **386** (2018) 012010 doi:10.1088/1757-899X/386/1/012010

- Analysis of regulations of the Russian Federation (RF) in relation to the requirements to the vehicle dynamics. Analytical calculation of the properties of vehicle dynamics. Definition of the optimization range
- Calculation of fuel economy properties for vehicle movement according to standardized cycles
- Verification and validation of calculation results using comparison with experiment data.
- Development of an algorithm and a program of the optimal gear ratios calculating. Optimization of gear ratios.

The article will describe in details the solution of problems 1-4. Verification and validation of calculations, as well as the development of an algorithm and a program of optimal gear ratios calculating, are planned at the next stages of the study.

2. Preparation of the initial data

The main part of the parameters for making calculations was determined by reference data or by information from the component manufacturer. These parameters include the current values of gear ratios, the transmission efficiency, the dynamic radius of the wheel, the cross-sectional area of the car and others. Using experimental studies, the specified ICE data were determined: the mechanical map - the dependence of the torque on the engine's rotational speed and on the fuel supply, the fuel map - the dependence of the specific fuel consumption on the engine's speed and on the load. The obtained data are shown in Figure 1.

Numerical values on the ordinate axis on the graphs are absent and this is due to the fact that this information is a commercial secret of the engine manufacturer. It should also be noted that for the prediction of harmful substances emissions values, also experimental data of ICE tests, namely the dependence of emission specific values on engine speed and its load is needed. At the moment, such tests have not been carried out, but this is planned for the subsequent stages of the study.

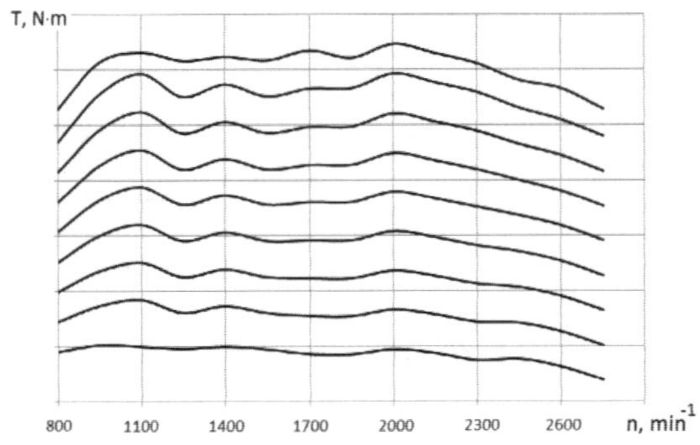

a)

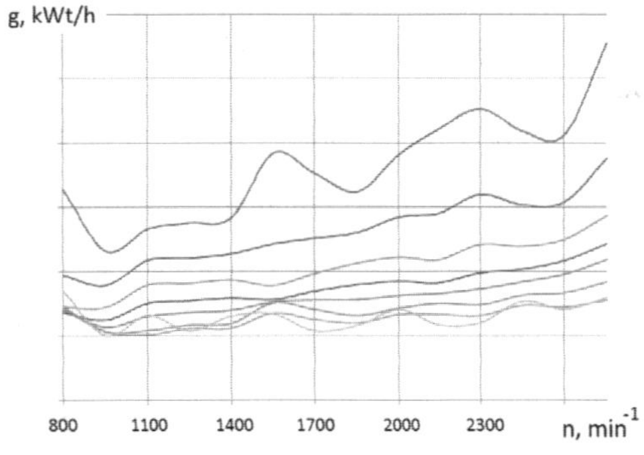

b)

Figure 1. Engine dynamic tests data:
a) a mechanical map, b) a fuel map

3. Analysis of single indicators of vehicle dynamics

In the Russian Federation, the next indicators of the vehicle dynamic properties are standardized by the technical regulations:

- Maximum speed. The test conditions are standardized by GOST 22576-90 [12]. The maximum permissible values are specified in GOST R 52280-2004 [13].
- Acceleration time on a specified distance (GOST 22576-90).
- Acceleration time to the set speed (GOST 22576-90).
- Speed characteristic "acceleration - run-out" (acceleration on the distance of 2000 m and deceleration (engine braking) to a stop - GOST 22576-90).

- Acceleration on the gear, providing the maximum speed (GOST 22576-90).
- Maximum slope (GOST 52280-2004).
- Speed on the long slope (GOST R 51734-2001 [14]).

For all these indicators, only the experimental determination conditions are normalized, and only for the first, sixth and seventh the permissible values are standardized. The first indicator sets the optimization limits for the top gear of the transmission or, taking into account the capabilities of the ICE, uniquely determines its value. The sixth and seventh indicators uniquely determine the minimum value of the first gear ratio and a gear on which the vehicle can move at a speed of at least 20 km/h on a 12% slope (most often the I or II gear). It should be noted that the last two indicators are relevant only for vehicles that are operated in mountainous terrain. All of the abovementioned indicators of vehicle dynamic properties can be determined using the formulas of the classical theory of the automobile [15]. Thus, it is possible to tell with confidence that it is desirable to optimize gear ratios 2 (in the limited range), 3, 4 and 5 numbers of transfer in case of the automobile gearbox with six speeds. If there is direct gear in the transmission (most often at the IV gear), this further narrows the optimization range, since direct transmission has a higher efficiency and a lower noise level. So the ratio of the direct gear can be changed only by correcting the main gear ratio value.

To take into consideration the dynamics of vehicle in optimization of the gear ratios, it is necessary to be guided by the conditions of determination of the parameter "acceleration time to the set speed" according GOST 22576-90.

4. Calculation of fuel economy for vehicle movement according to standardized cycles

For assessing the fuel efficiency of cars according GOST R 54810-2011 [16], two types of driving cycles have been established: highway and urban.

Simulation of the vehicle's movement is performed in accordance with technological (operational) maps and traffic patterns in the highway cycle (Qs hc) and urban cycle (Qs uc) on a straight and horizontal road.

The main provisions (assumptions) of the methodology are formulated in the following form:

- a continuous process of a car motion can be represented by a combination of individual components of the movement: acceleration from zero and non zero velocity; constant speed; engine braking; motion when the engine is idling;
- dynamic motion of a car (the acceleration is not constant) can be represented by a combination of modes with pre-determined variations in the speed of motion, within which the acceleration remains constant;
- partial load characteristics of the engine can be used in calculation of the fuel economy properties of the car.

The adopted assumptions make it possible to substantially simplify the problem of calculating fuel consumption for an uneven motion. Modeling of the vehicle motion when the engine is running at partial loads allows to differentiate the processes of unsteady motion and calculate the actual characteristics of fuel economy for small values speed change intervals. The made assumptions are expected to provide accuracy in fuel consumption calculating, not only in motion according standardized cycles, but also in motion along a real route the structure of which is determined by the specified components.

In simulation of motion in driving cycles, all the requirements of regulatory documents regarding the choice of gears for acceleration and motion, fuel supply during acceleration, requirements for possible speed deviations from the set value are taken into account.

The scheme of the highway cycle consists of next pieces:

- acceleration starting from the constant non zero velocity;
- constant speed motion;
- engine braking.

The scheme of the urban cycle is distinguished by the presence of next types of motion:

- acceleration starting from zero speed;

- service braking with deceleration of 1m/sec²;
- engine work in an idling mode.

The urban cycle is fundamentally different from the highway cycle by the presence of acceleration starting from zero speed. As acceleration is started from the lowest transfer and carries out in the rated power of the engine, acceleration of the vehicle can't remain constant, and the fuel consumption should be determined according to the specified circumstances. The second feature of the cycle is the presence of service braking up to vehicle stopping and the engine is idle. According GOST 54810 - 2011, service braking is performed on a given distance (the path length is determined from the average deceleration of 1 m/sec²) with the gear engaged using the service brakes. When the vehicle brakes to a full stop, the gear should be switched to the neutral position when the minimum speed is reached.

Description of the standardized cycles does not determine either the distance or the time of accelerations.

Uncertainty of the requirements of regulatory documents with respect to the parameters of motion during acceleration require additional computational operations that allow us to establish the dependencies or graphs of the characteristics (functions) a = f (V) and V = f (S) obtained under the conditions of tests with full fuel supply.

As a result of preliminary calculations, it was found that the accelerations according requirements of regulations and on gear number, used at the beginning and end of the acceleration paths, vary insignificantly, by approximately 8...12%. This circumstance allowed us to determine the mean values of the accelerations for each of the acceleration distances and in subsequently determine the fuel consumption for the uniformly accelerated motion.

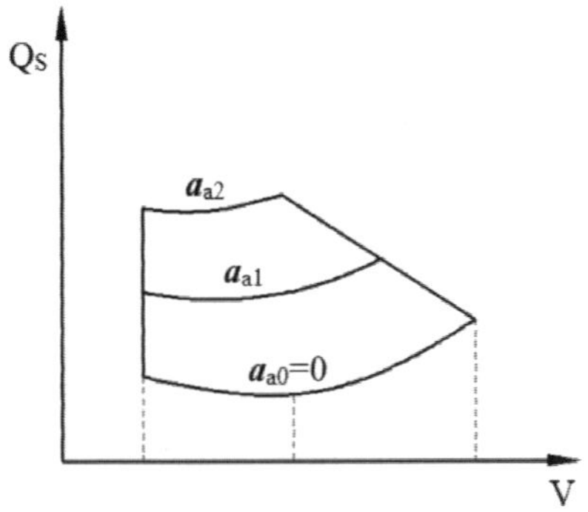

Figure 2. – Determination of fuel consumption in the acceleration distance

From this moment, the assumption about the linearity of the acceleration characteristics is introduced, due to insignificant variations of the acceleration values at each moment of time. The above assumption substantially simplifies the task of calculating of fuel consumption at acceleration distances due to the possibility of using the fuel characteristics of vehicle acceleration. To determine the fuel consumption on the sections of motion at a constant speed, the fuel consumption characteristics of the steady-state traffic for the respective gears are determined by the cycle description. Fuel consumption characteristics link the vehicle's design parameters and motion conditions. The graph of the characteristic represents the dependence of the fuel consumption Qs on the steady-state velocity (V = const) on roads with different values of the drag coefficient ψ (figure 2). Assuming that the car is moving without slipping driving wheels, evenly, without acceleration (a = 0) and the power expended for acceleration is equal to Pa = 0, the fuel consumption is calculated by formula [15]:

$$Q_s = \frac{g_e(P_\psi + P_v)}{36V\rho_t k_c \eta_{tr}}, \tag{1}$$

where g_e – specific fuel consumption; ρ_t – fuel density; η_{tr} – efficiency of transmission; k_c – coefficient of correction of power and torque values obtained using bench tests; V – velocity; Pψ, Pv – respectively, the powers expended to overcome the resistance of movement and air resistance.

The fuel consumption characteristics of acceleration Qs = f (V, ψ), fig. 2, can be calculated using the fuel consumption characteristic of steady motion. Assuming if the fuel consumption is calculated when driving on the road with a constant value of the coefficient ψ_1 (with acceleration a = 0), then the motion along the road with the drag coefficient $\psi_2 < \psi_1$, other things being equal, can be considered as uniformly accelerated, at which the resistance to acceleration will be equal to:

$$F_{a1} = \delta m_a a_{a1} = F_{\psi 2} - F_{\psi 1}, \tag{2}$$

where Fa_I – acceleration resistance force; δ – coefficient of acceleration of rotating masses; a_{aI} – vehicle acceleration; Fψ1, Fψ2 – forces of resistance to movement, respectively, for resistance coefficients ψ1 and ψ2.

If the characteristic of the fuel flow rate is specified by the drag coefficient ψ_n, Qs = f (V, ψ_n) then the value of the constant acceleration of a car whose engine is running in full power mode can be determined from expression:

$$a_{an} = \frac{F_{\psi n}-F_{\psi 1}}{\delta m_a}, \tag{3}$$

where a_{an} – vehicle acceleration at motion on the road with a drag coefficient ψ_n.

The resistance force at uniformly accelerated motion on the acceleration distances is determined according expression::

$$F\psi_I = Fa_I + F\psi_1 = \delta m_a\, a_I + F_{\psi 1}, \tag{4}$$

where FψI – resistance force at uniformly accelerated motion on the acceleration distances of the highway cycle; a_I –average value of the acceleration on the acceleration distance; $F\psi_1$ – resistance force at uniform movement of the vehicle on the road with a drag coefficient ψ_1; m_a – gross vehicle mass.

To calculate the fuel consumption in the acceleration distances of the cycles, the fuel consumption characteristics of the vehicle Qs = f (V, a_I) for the average acceleration values a_I in the acceleration distances should be determined. The abscissa (speed) within each part of the acceleration must be divided into sufficiently small intervals ΔVi such that the initial ΔVi-1 and the final value ΔVi of the velocity within the selected segment ensure minimum deviations from the mean value of the speed of the segment: The distance passed by the vehicle during the accelerating time provided that acceleration constant, and average speed in each selected segment in the accelerating section - equal t:

$$V_{i\,av} = \frac{V_i+V_{i-1}}{2}, \tag{5}$$

where Vi av – average speed of the vehicle within the selected segment; Vi-1, Vi – respectively, the initial and final values of the velocities of the selected segments.

The time for overcoming the selected segments is defined as

$$\Delta t_i = \frac{V_i-V_{i-1}}{a_I}, \tag{6}$$

where Δt_i – time for overcoming the selected segment; a_I – acceleration.

The distance traveled by a car within the selected segment of velocities, taking into account expressions (5) and (6), is calculated by formula:

$$\Delta S_i = V_{i\,av}\, \Delta t_i = \frac{V_i^2-V_{i-1}^2}{2a_I}, \tag{7}$$

where ΔSi - distance traveled by a car within the selected segment.

The fuel volume Qi (l) consumed on the segment V_{i-1}-V_i is determined according expression:

$$Q_i = \frac{Q_{si}\Delta S_i}{100}, \tag{8}$$

where Qsi – fuel consumption on the segment of the speed change V_{i-1} - V_i; Q_i – volume of fuel consumed on a segment V_{i-1} - V_i.

The volume of fuel consumed at the i-th acceleration section can be calculated by the formula:

$$Q_{acc.\ i} = \sum_{i=1}^{k} \frac{Q_{si}\,\Delta S_i}{100} = \frac{1}{200a_I}\sum_{i=1}^{k} Q_{si}(V_i - V_{i-1}), \tag{9}$$

where $Q_{acc.i}$ - fuel volume consumed in the acceleration section; k – the number of intervals (segments) in the range of change in speed during acceleration; V_i – initial speed of the acceleration section; V_k – final speed of the acceleration section.

When calculating amount of fuel consumed by a vehicle on sections of a driving cycle with engine braking, the section is divided into segments ΔS_j such that within the segment the car's deceleration can be considered as constant. Taking into account the expression for the average speed of movement, the average angular speed of the engine's crankshaft can be set for the each segment

$$w_{ei} = \frac{V_{j\,av}u_{tr}}{r_k}, \tag{10}$$

IOP Conf. Series: Materials Science and Engineering **386** (2018) 012010 doi:10.1088/1757-899X/386/1/012010

where w_{ei} - average angular rotation frequency of a crankshaft; V_{jav} - average speed; u_{tr} – transmission ratio; r_k – rolling radius of a wheel.

The load power of the engine P_{ei} at motion with engine braking is determined in accordance with expression:

$$P_{e\,i} = \frac{P_{\psi i}+P_{vi}}{k_c\eta_{тр}} + P_{fr.}, \tag{11}$$

where $P_{\psi i}$, P_{vi} – respectively, the power expended to overcome the resistance of the road and resistance of the air; $P_{fr.}$ – power expended to overcome the friction in the engine.

In the event that the friction torque $T_{fr.0}$ in the engine is known at a minimum angular velocity, the value of the motor power loss $P_{fr.}$ can be calculated at any travel speed.

$$P_{fr.} = \frac{(T_{fr.0}+b_t u_{tr} V_{j\,av}) u_{tr} V_{j\,av}}{1000 r_k^2}, \tag{12}$$

where b_t - coefficient of torque accounting the change in the crankshaft rotation speed.

In the case the component $P_{fr.}$ of the equation (11) can not be determined, the engine load power in the engine braking mode is determined taking into account the first term.

Overall fuel consumption during engine braking in the section ΔS_i is determined by means of the known dependence and the total power loss in this case is equal to the power of losses in the engine.

$$Q_{sj} = \frac{g_{ej} P_{fr.j}}{36 V_{j\,av} \rho_f k_c \eta_{tr}} \tag{13}$$

The amount of fuel consumed in the deceleration area is determined in accordance with expression (9).

The amount of fuel consumed when the engine is idling Q_{im} can be determined using expression

$$Q_{im.} = \frac{G_f t}{\rho_f}, \tag{14}$$

where G_f – hourly fuel consumption; t – the time of engine idling.

The engine operating time in this mode of motion must be determined taking into account the operating card of the cycle.

The fuel consumption in the areas of service braking is determined taking into account the features of using the service brake system. When braking using the service braking system, the engine operates in idling mode. Therefore, to calculate the fuel consumption in this mode of motion according formula (14), it is sufficient to determine the service braking time according the operating card of the cycle.

The total fuel consumption in the cycle is composed of all components of the cycle scheme: acceleration, constant speed, engine braking, engine idling and service braking.

For example, the fuel consumption Qh in the highway cycle, taking into account all the components of the motion, is given by

$$Q_h = \sum_{I=1}^{3} \sum_{i=1}^{k} \frac{Q_{si}\,\Delta S_i}{100} + \sum_{J=1}^{3} \sum_{j=1}^{m} \frac{Q_{sj}\Delta S_j}{100} + \sum_{e=1}^{6} \frac{Q_{se} S_e}{100}, \tag{15}$$

where i – number of acceleration sections; j – number of engine braking sections; m – number of selected segments in the engine braking section; e – number of traffic sections with constant speed; Se – length of a section of motion with constant speed. The motion schemes and graphs of the current fuel consumption calculated according to the above mentioned formulas, respectively for the highway and urban cycles are shown in figures 3-6.

IOP Conf. Series: Materials Science and Engineering **386** (2018) 012010 doi:10.1088/1757-899X/386/1/012010

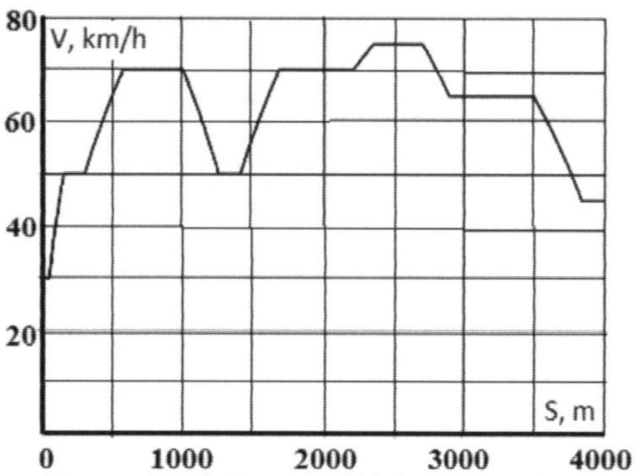

Figure 3 - Scheme of highway cycle

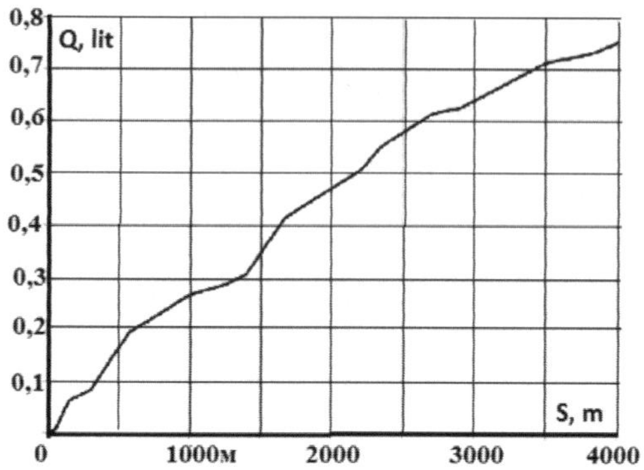

Figure 4 – Curve of current fuel consumption in the highway cycle

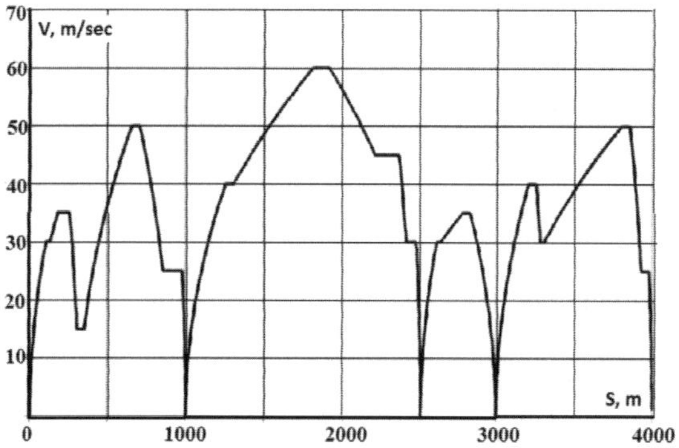

Figure 5 – Scheme of urban cycle

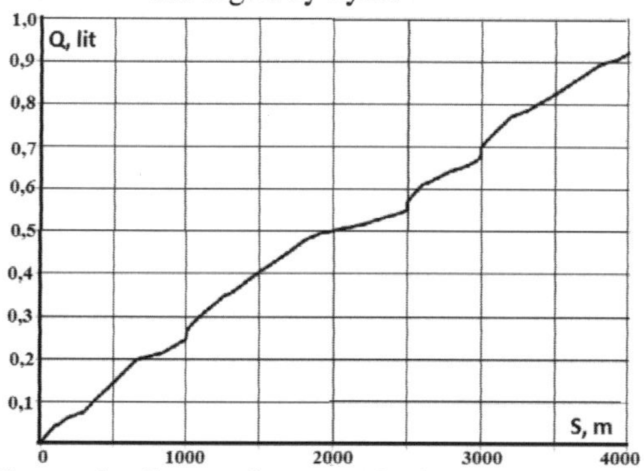

Figure 6 - Curve of current fuel consumption in the urban cycle

5. Conclusions

- The analysis of the works in the field of transmission gear ratios optimization is carried out. The most rational approaches to the solution of the problem of transmission gear ratios optimization was identified from the point of view of a simultaneous increasing of vehicle dynamics, reducing of fuel consumption and reducing of emissions of harmful substances.
- The initial data were prepared to perform the calculations. During the experiment on the dynamometer bench, a mechanical and fuel map of the ICE were obtained. The dependence of the specific values of the emissions on the speed of internal combustion engine and the load on the engine is expected to be obtained at the next stages of the study.
- The analysis of the current regulations of the Russian Federation in relation to the requirements to the indicators of vehicle dynamics properties is carried out. The numbers of gears are determined, the ratio of which is uniquely determined by the conditions for the fulfilment of the regulations. Calculation methods for single indicators of the properties of vehicle dynamics are determined.
- A methodology for the analytical calculation of fuel consumption in motion according standardized cycles has been developed and demonstrated. The obtained results of calculation according developed technique are shown.
- The next steps that are planned to be done to complete the study are verification and validation of the calculation results by comparisons with field tests results and developing an algorithm and a program for calculating the optimal gear ratios of the vehicle.

IOP Conf. Series: Materials Science and Engineering **386** (2018) 012010 doi:10.1088/1757-899X/386/1/012010

This research was conducted with the financial support from Ministry of Education and Science of the Russian Federation in the frame of the complex project "The establishment of the high-tech manufacturing of safe and export-oriented GAZ vehicles with autonomous control systems and the possibility of integration with the electric platform on the base of components of Russian production" under the contract №03.G25.31.0270 from 29.05.2017 (Governmental Regulation №218 from 09.04.2010).

References

[1] Uniform provisions concerning the approval of vehicles with regard to the emission of pollutants according to engine fuel requirements. Date Views 01.05.2018 www.unece.org/fileadmin/DAM/trans/main/wp29/wp29regs/R083r5e.pdf.

[2] Uniform provisions concerning the measures to be taken against the emission of gaseous and particulate pollutants from compression ignition engines and positive ignition engines for use in vehicles. Date Views 01.05.2018 www.unece.org/fileadmin/DAM/trans/main/wp29/wp29regs/2013/R049r6e.pdf.

[3] Zhao X X, Ing A H., Azad N. l., McPhee J 2015 An optimal gear-shifting strategy for heavy trucks with trade-off study between trip time and fuel economy. International Journal of Heavy Vehicle Systems. 22(4). pp 356–374. https://doi.org/10.1504/IJHVS.2015.073205

[4] Robinette D and Wehrwein D 2015 Automatic transmission gear ratio optimization and Monte Carlo simulation of fuel consumption with parasitic loss uncertainty. SAE Int J. Commer. Veh. 8(1). doi:10.4271/2015-01-1145

[5] İlker Küçükoğlu 2016 Differential evolution algorithm for gear ratio optimization of vehicles. Int. Journal of Engineering Research and Application. 6(10) Part 2. pp.29-33

[6] Jammalamadaka S, Schaeflein P and Kauffmann P 2018. Impact of BEV powertrain architectures on energy consumption in various driving cycles. Date Views 20.04.2018 www.gtisoft.com/wp-content/uploads/2018/01/GT-Suite-Conference_V4.pdf

[7] Jixiang F, Jiangyan Z, Tielong S 2015. Map-based power-split strategy design with predictive performance optimization for parallel hybrid electric vehicles. Energies. 8. pp. 9946-9968

[8] Goharimanesh M, Akbari A, Tootoonchi A A 2014. More efficiency in fuel consumption using gearbox optimization based on Taguchi method. J. Ind. Eng. Int. 10(61). DOI 10.1007/s40092-014-0061-y

[9] Ngo D V, Colin Navarrete J A, Hofman T, Steinbuch M,. Serrarens A F A 2013 Optimal gear shift strategies for fuel economy and driveability. Proceedings of the Institution of Mechanical Engineers. Part D : Journal of Automobile Engineering. 10(227). pp. 1398-1413

[10] Reghunath S K, Sharma D, Athreya A S 2014 Optimal gearshift strategy using predictive algorithm for fuel economy improvement. SAE Technical Paper 2014-01-1743, doi: 10.4271/2014-01-1743

[11] Wang W, Chen C 2013 Sport Utility Vehicle Refinement by Engine Upgrading and Gear Ratio Optimization. International Journal of Advanced Science and Technology. 54. pp 61-76

[12] GOST 22576-90 Avtotransportnye sredstva. Skorostnye svojstva. Metody ispytanij. Date Views 20.04.2018 www.internet-law.ru/gosts/gost/28319/

[13] GOST R 52280-2004 Avtomobili gruzovye. Obshchie tekhnicheskie trebovaniya. Date Views 20.04.2018 www.internet-law.ru/gosts/gost/4773/

[14] GOST R 51734-2001 Avtotransportnye sredstva dlya gornyh marshrutov. Tekhnicheskie trebovaniya. Date Views 20.04.2018 www.internet-law.ru/gosts/gost/6894/

[15] Kravec V N 2013 Teoriya avtomobilya: uchebnoe posobie - 2 izd. Nizhegorod. gos. Tekhn. Un-t im. R.E. Alekseeva. 413 p.

[16] GOST R 54810-2011 Avtomobil'nye transportnye sredstva. Toplivnaya ehkonomichnost'. Metody ispytanij. Date Views 25.04.2018 www.internet-law.ru/gosts/gost/51790/.

IOP Conf. Series: Materials Science and Engineering **386** (2018) 012011 doi:10.1088/1757-899X/386/1/012011

Development of technical solutions for development of a commercial vehicle driver assistance systems

D M Porubov[1], A M Groshev[1], V P Mishustov[1], Y I Palutin[2], A V Tumasov[1]

[1]Nizhny Novgorod State Technical University n.a. R.E. Alekseev, Minin str., 24, Nizhniy Novgorod, 603950, Russian Federation
[2] Nizhny Novgorod State Agriculture Academy

E-mail: pavel.beresnev@nntu.ru

Abstract. As the world's stock of cars is growing, the number of road accidents per each 100000th resident is going up as well. To avoid road accidents or to mitigate their consequences nowadays driver assistance systems are used. Pre-requisites and first steps of driver assistant systems development and integration in series-produced vehicles are presented in this article. Review of current solutions is given and technical regulation development trends in this field are considered. An option of driver assistance systems integration on the platform of GAZelle NEXT is offered. It has also been detected that current series-produced GAZelle NEXT vehicles provide no opportunity of external control over actuating devices. This factor hinders development of domestically made vehicles and reduces their export potential. Force and kinematic analysis of the steering system is presented in MSC ADAMS/View program package. Also a number of system designs is presented for further mounting on "GAZ Group" vehicle family, having a feature set of recognizing vehicles, road signs, pedestrians, road marking and traffic lights; warning of possible head-on clash and leaving the traffic lane. Field research of the developed driver assistant system has been carried out. Ways of further integration of driver assistant systems in series-produced "GAZ Group" vehicles are mapped out.

1. Introduction

Transport plays an important role in forming and functioning of the modern society. It helps to provide equal access to different objects and services. Transport and transport communications are one of the key factors providing development of industrial complexes, defence and state security. In this regard transport communications help to raise information awareness, stimulate science and innovations, economic activity, global character of different human activities including the country's northern territories development.

A developed transport network ensures connection between all the country's territories making them a single social, cultural and economic space. As soon as the Russian Federation occupies the first place in the world in terms of its territory, this is the country's crucial parameter and a necessary condition of its territorial integrity.

Improving road traffic safety remains one of the basic conditions of transport network development in the Russian Federation. According to the Russian government statistics each year 20 deaths per 100 000 residents are registered in road accidents on average. In the USA, this number is 10-12 deaths per 100 000 residents and to Germany – 5-6 deaths. A relatively big number of road accidents involve commercial vehicles due to their operational specialty (minimal, simple, a big number of transport operations, haul length etc.). Herewith, one of the main causes of road accidents in Russia is the so-called «human factor» which is also characteristic of other world community countries. For example, in Europe over 90% of road accidents happen because of the drivers' mistakes, in particular because drivers and motorists are distracted from the process of driving.

Research and development of driver assistance systems is carried out in order to minimize drivers' mistakes and to improve road traffic safety for commercial vehicles inclusively. The basic algorithm of ADAS-systems operation is collection of sensor data around the moving vehicle , further information analysis, detecting emergency situations on the road and taking measures to avoid them or to mitigate their consequences. Nowadays a number of measures is being taken to alter the regulatory framework and implement the requirements to have series-produced vehicles equipped with driver assistance systems. These alternations encourage a breakthrough development of driver assistance systems as well as their further integration in series-produced vehicles in order to improve road traffic safety.

2. Technical regulation in driver assistance systems

Technical regulation is considered to equip commercial vehicles of N1, N2- categories with different driver assistance systems. Implementation of driver assistance systems in commercial vehicles is an essential task as it improves road traffic safety due to these vehicles operational specialty (minimal idle time, haul length etc.) and a big driver's workload, especially when handling regular cargo-and-passenger hauls.

In the European Union countries nowadays there are requirements to equip vehicles of N2- category with driver assistance systems. Currently the following systems are obligatory: Automatic Emergency Braking, Lane Departure Warning; Automatic headlight control system.

Starting from 2022 new EU requirements are to be enacted to equip N1-category vehicles with such ADAS types as Automatic Emergency Braking (AEB), Emergency Braking Display; Intelligent Speed Adaptation; Safety Belt Reminders; Alcohol Interlock Devices; Crash Event Data Recorder; *Tyre pressure Monitoring*; Reversing Detection, Lane Keep Assistance; Driver Drowness and Distraction monitoring [1,2].

According to the EU requirements to equip N1-category vehicles starting from 2024, the following ADAS will be obligatory: Automatic Emergency Braking (AEB), Lane Keep Assistance; Driver Drowness and Distraction monitoring.

Nowadays in the Russian Federation there are no obligatory requirements to equip vehicles of N1, N2- categories with driver assistance systems. However implementation of the requirements analogous to UNECE Rules is planned in normative legal documents of the Russian Federation, videlicet in Technical Regulation of Customs Union TRCU 018/2011 «On wheeled vehicles safety». For example, implementation of Automatic Emergency Braking (AEB) requirements is due in the period of 2019-2021 [3]. Analogous requirements are offered with respect to Forward Collision Warning System, road signs and traffic lights detection system, lane leaving system.

In Russia production and sales leader in LCV segment (about 70% of the market) is "GAZ Group" LLC [4]. The basic model in this class is GAZelle NEXT. Different design variants of the vehicle (cargo and passenger) are assembled on the basis of this chassis type. Nowadays this model has no driver assistance systems installed. However to keep and increase the market share in Russia as well as to increase "GAZ Group" production export potential it is necessary to install driver assistance systems complying with both UNECE and Russia's normative legal requirements and meeting the needs of target customers.

3. Review of current solutions

As soon as normative requirements to equip commercial vehicles with driver assistance systems will first of all concern N1-category, light commercial vehicles manufacturers have already started to equip series-produced vehicles with different ADAS types. Benchmarking has been carried out in order to assess level of development of driver assistance systems installed in series-produced commercial vehicles of N1-category. The obtained data are shown in Table 1

IOP Conf. Series: Materials Science and Engineering **386** (2018) 012011 doi:10.1088/1757-899X/386/1/012011

Table 1. Benchmarking data

	IVECO Daily [5]	Ford Transit[6]	Volkswagen Crafter[7]	Mercedes-Benz Sprinter[8]	Peugeot Boxer[9]	Hyundai H350[10]
Adaptive Cruise Control		+	+			
Automatic Emergency Braking (Automatic Brake Assist)			+	+		
Blind spot monitor			+	+		
Collision avoidance system (Precrash system)			+	+		
Forward Collision Warning		+	+	+		
Lane change assistance			+			
Lane Keep Assist			+			
Hill Start Assist		+		+	+	+
Cruise Control	+			+	+	+
Downhill Asssist Control				+	+	
Parking Assistance	+	+	+	+	+	+
Driver Monitoring System		+	+			
Navigation System		+	+	+		+
Automatic Emergency Braking system				+	+	+

According to the table such carmakers as Volkswagen (car model "Crafter") and Mercedes Benz (car model "Sprinter") [7, 8] enjoy the highest rate of using ADAS in LCV. Basically the above commercial vehicles are equipped with the following systems: Hill Start Assist, Cruise Control, Downhill Assist Control, Parking Assistance и Automatic Emergency Braking system.

Only Volkswagen, Iveco and Mercedes have active driver assistance systems of the second level according to SAE classification (affecting driving control units) [11]. Consequently, these systems allow the companies to give their competitors an edge in this market segment as well as to increase the rate of development of breakthrough technologies in the car industry. and driver assistance systems in particular.

However it is worth mentioning that in LCV segment renewal takes place once in 5 - 10 years. In the nearest future other global carmakers are likely to equip their vehicles with active driver assistance systems of such level, seeing their competitors produce next generation LCVs.

Analysis of normative documents of commercial vehicles market shows that development and equipment of "GAZ Group" with driver assistance systems, namely GAZelle NEXT, will help largely increase the safety level of competitive production as well as to expand export potential of commercial vehicles.

4. Driver assistance systems integration

To actuate control actions from the driver assistance system control unit it is necessary to provide external control of principle actuating mechanisms of the vehicle. At present series-produced GAZelle NEXT vehicles have no possibility of external control of actuating mechanisms, namely brake system and steering control system.

Within the framework of this project intelligent brake system and steering control system with the possibility of external control are being developed.

Intelligent brake system with the possibility of external control will serve as an actuating mechanism for Automatic Emergency Braking system and will be able to respond to signals coming from electronic control unit of the vehicle showing the risk of collision with the vehicle in front. It will also activate braking mechanisms in order to slow down to prevent the collision or mitigate its consequences. At this, the system will give the driver the possibility to interrupt the emergency braking stage and will provide operation within the range of 15 km/h to the maximum speed of the vehicle except for the cases when it

IOP Conf. Series: Materials Science and Engineering **386** (2018) 012011 doi:10.1088/1757-899X/386/1/012011

is manually switched off. At this stage of the project the structural components of the system have been developed and assembly operations have been carried out. Fig. 1 shows the components layout of the brake system under development in GAZelle NEXT vehicles.

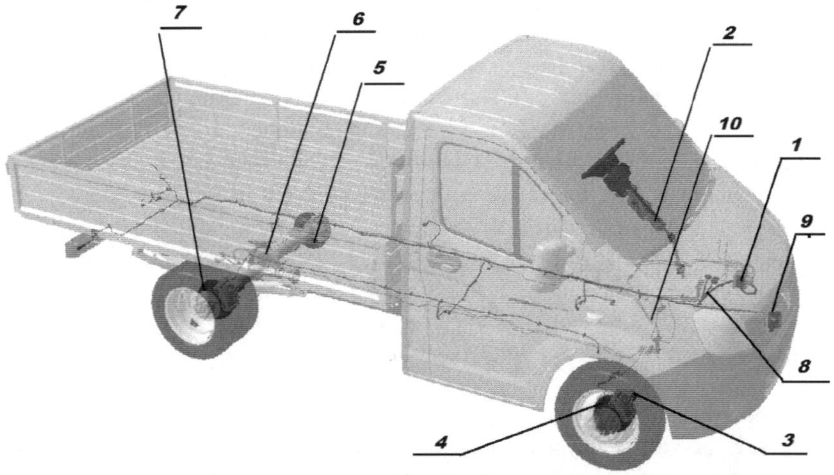

Fig 1. Intelligent brake system in GAZelle NEXT vehicles: 1 – hydraulic unit with external control function; 2 – steering wheel angle sensor; 3 – ABS sensor; 4 – front brake rotors; 5 – rear brake rotors; 6 – drive incorporating a vacuum booster; 7 – park brake; 8 – electric circuit; 9 – radar with an integrated control unit; 10 – brake system.

When developing the steering control system an analysis has been carried out of the current steering control systems used in series-produced vehicles equipped with driver assistance systems, directly influencing the steering control. The results of the analysis allowed to come to a conclusion that in order to successfully complete the above tasks it is necessary to use the steering control system with electromechanical power steering.

On the first design stage technical specifications for the steering control have been determined, the structural components of the system have been developed and assembly operations have been carried out. Also on this design stage force and kinematic analysis of the steering control has been carried out using MSC ADAMS/View program package.

In the course of modelling turning of the wheels depending on the suspension vertical travel as well as the wheel turning kinematics were researched. Fig. 2 shows general view and model (MSC ADAMS) of GAZelle NEXT vehicle front suspension with steering control under development.

IOP Conf. Series: Materials Science and Engineering **386** (2018) 012011 doi:10.1088/1757-899X/386/1/012011

a б

Fig. 2. (a) General view of steering control; (б) Steering control model

With the modelling results in view a comparative assessment of the steering control under development and the current series-produced steering control with a hydraulic booster was carried out.
Kinematic analysis has shown that the amplitude of changing toes at the suspension vertical travel with the steering rack under development and a standard steering mechanism is similar. The amplitude of suspension with a standard steering mechanism was 0,031°, the one with the steering mechanism under development was 0,028°. Because of a smaller steering ratio and a bigger stroke of the steering mechanism under development, maximum wheel turning angles are larger than when using standard steering mechanism. Also this model of suspension with the steering control helped to determine a boosting ratio for the electric power steering. As a result, the electric motor torque dependence upon the steering wheel torque has been obtained. Basing on the modelling results it is possible to make a conclusion about acceptable force and kinematic specifications of the steering control under development. The modelling results will be further used to calculate electric motor specifications and to develop algorithms of the control system electronic unit operation.

5. Developing driver assistance systems

Alongside with developing brake and steering systems with outer control, operations are carried out to seek technical solutions for driver assistance systems. Basing on the former presented analysis of driver assistance systems for light commercial vehicles, an idea has been put forward to develop a number of systems for further mounting on "GAZ Group" vehicle family, with a specific feature for recognizing vehicles, road signs, pedestrians, road marking and traffic lights; warning of possible head-on clash and leaving the traffic lane.
The structural components of the traffic objects recognition systems and warning of possible head-on clash as well as the system of leaving the traffic lane are represented by calculation research module Nvidia Jetson TX2 and optical sensor Basler acA1920-50gc.
Field research of the system efficiency at recognizing road signs and road users with warning of possible crashes was carried out in city traffic with various traffic flow and weather conditions. It should be noted that nowadays the system is able to recognize a number of road signs having a great influence on the safety of traffic (pedestrian crossing, main road, STOP sign, speed limit, etc.). In future the functional of the system will be completed. A fragment of the system test in winter in high humidity is shown in Fig. 3. The system marks the vehicle in the way with red.

a б

Fig 3. (a) recognition of vehicles moving along open roads;
(б) the system recognition of road signs

IOP Conf. Series: Materials Science and Engineering **386** (2018) 012011 doi:10.1088/1757-899X/386/1/012011

The system tests of head-on clash warning were carried out on a special restricted test area. A GAZelle NEXT vehicle having a system of crash warning was tested. The system algorithm is the following: The system recognizes the object in the way by a video camera and estimates the distance to it. Using travelling speed data from CAN-bus of the vehicle the approach speed to the object depending on safety distance to it is estimated at every moment. In case of exceeding the permissible limit, the driver gets a warning signal of a possible crash. It should be noted that the brakeage took place after the signal started. A fragment of successful brakeage in front of the pedestrian manikin is shown in Fig. 4.

Fig. 4. Successful brakeage in front of the manikin.

The criterion of correct system actuation was a timely signal warning the driver of possible clash with a pedestrian. The calculated accuracy of critical situation estimate was 0,95. The completeness showing the number of timely actuations of the total number of tests was 0,94. The system performance index was 95%.
The warning system operation when leaving the traffic lane is based on video camera data. Next, according to the system operation algorithm, the video image is processed and the system determines the value of deviation from the center of the lane.
The algorithm and evaluation of correct system actuation are shown in [12] There are 3 warning areas: the green one – the vehicle is in the center of the lane (Fig. 5a); the yellow one – deviation from the center of the lane within the range of 0.2 to 0.3 m (Fig. 5б); the red one – deviation from the center of the lane of more than 0.3 m (Fig. 5в). The yellow area is transitional, showing visual information when the driver finds himself in a possible emergency situation. When the red area is displayed, the system warns the driver of an emergency situation by an audio signal. A fragment of the system operation is shown in Fig. 5.

a б в

Fig 5. A fragment of the warning system operation when leaving the traffic lane:
(a) The green area, (б) – The yellow area (deviation of 0.2-0.3 m.), (в) – The red area (deviation of more than 0.3 m.)

6. Conclusion

Increasing the traffic safety in the nearest future by implementing driver assistance systems will be an obligatory requirement for car makers according to the analysis of UN ECE normative documents and technical rules of the Russian Federation. Basing on market research of LCV, equipped with ADAS, it is possible to make a conclusion that some car makers are ready to equip vehicles with obligatory driver assistance systems in series.

To keep the market share of the Russian Federation and increase the export potential, the research team of NNSTU after R.E. Alexeev is working together with engineers of Joint Engineering Center of "GAZ" Group to prepare vehicles of LCV category for implementation of driver assistance systems, as well as direct development of ADAS.
Following the project the structural components of intelligent brake system and steering control system with outer control have been developed and assembly operations have been carried out. When developing the steering control force and kinematic analysis has been carried out. Basing on the modelling results it is possible to make a conclusion about acceptable force and kinematic specifications of the steering control under development. The modelling results will be further used to calculate electric motor specifications and to develop algorithms of the control system electronic unit operation.
Also driver assistance systems have been developed and tested with a special feature for recognizing vehicles, road signs, pedestrians, road marking and traffic lights; warning of possible head-on clash and leaving the traffic lane. According to the test the system of vehicle recognition has shown good results, "missing"
on-coming, moving in the same direction and parked vehicles only in rare cases. A number of pre-set road signs are recognized satisfactorily but neural network training is necessary to recognize a bigger number of road signs.
Test results have shown that the system of obstacles recognition is efficient, but it can become more efficient when active sensors (radars) to estimate the distance to the obstacle are used. In future when operating intelligent brake system it will be become possible to set brake control action. This solution will automatically provide emergency braking in front of the obstacle thus largely increasing traffic safety.
Test results have shown that the system warning when leaving the traffic lane functions satisfactorily. However, in future it is necessary to consider completing this system with synchronization with turn indicators of the vehicle, the function of tracking the driver's operating the steering wheel, i.e. the possibility to track if the driver's hands are on the steering wheel. Also after developing the steering system with external control it is necessary to realize the steering control action to prevent unintended leaving the traffic lane.
Realization and integration of actuating mechanisms with external control and driver assistance systems, meeting UN ECE legal requirements, Russian legislation and customer needs on commercial vehicles GAZelle NEXT will allow to significantly increase traffic safety as well as to keep the Russia's market share and boost export potential of "GAZ" Group.

Acknowledgments
This research has been carried out with the financial support from Ministry of Education and Science of the Russian Federation in the frame of the complex project "The establishment of the high-tech manufacturing of safe and export-oriented GAZ vehicles with autonomous control systems and the possibility of integration with the electric platform on the base of components of Russian production" under the contract №03.G25.31.0270 from 29.05.2017 (Governmental Regulation №218 from 09.04.2010).

References
[1] Benefit and Feasibility of a Range of New Technologies and Unregulated Measures in the fields of Vehicle Occupant Safety and Protection of Vulnerable Road Users / D Hynd, M McCarthy, J Carroll, M Seidl, M Edwards, C Visvikis, M Tress, N Reed, A Stevens, - Brussels: European Commission, 2015. - 470 c.
[2] Saving Lives: Boosting Car Safety in the EU. Reporting on the monitoring and assessment of advanced vehicle safety features, their cost effectiveness and feasibility for the review of the regulations on general vehicle safety and on the protection of pedestrians and other vulnerable road users / Comission Staff Working Document - Report from the Commission to the European Parliament and the Council, - Brussels: European Commission, 2016. - 56 c.

[3] Kupriyanovsky V., Akimov A., Pokusaev O., Alenkov V., Namiot D., Sinyagov S. Intellectual mobility and mobility as a service in Smart Cities //International Journal of Open Information Technologies. – 2017. – T. 5. – #. 12. – S. 77-122.
[4] GAZ Group. Retrieved May 30, 2018, from: http://gazgroup.ru/
[5] Iveco. Retrieved May 30, 2018, from: https://www.iveco.com/
[6] Ford. Retrieved May 30, 2018, from: https://www.ford.com/
[7] Volkswagen.Retrieved May 30, 2018, from: http://www.vw.com/
[8] Mercedes-benz. Retrieved May 30, 2018, from: https://www.mercedes-benz.com/
[9] Peugeot. Retrieved May 30, 2018, from: http://www.peugeot.com/en
[10] Hyundai. Retrieved May 30, 2018, from: https://www.hyundai.com/worldwide/en
[11] SAE. Retrieved May 30, 2018, from: https://www.sae.org/attend/adas/
[12] Beresnev, P., Tumasov, A., Tyugin, D., Zeziulin, D., Filatov, V. and Porubov, D.Automated Driving System based on Roadway and Traffic Conditions Monitoring. In Proceedings of the 4th International Conference on Vehicle Technology and Intelligent Transport Systems (VEHITS 2018), pages 363-370, ISBN: 978-989-758-293-6

IOP Conf. Series: Materials Science and Engineering **386** (2018) 012012 doi:10.1088/1757-899X/386/1/012012

The thermal loading estimation of the friction pairs of a vehicle automated brake system

V Dygalo[1], I Zhukov[2]

[1]D. Sc., Professor, Volgograd State Technical University, Russia, 400005, Volgograd, Lenin avenue, 28
[2]Graduate student, Volgograd State Technical University, Russia, 400005, Volgograd, Lenin avenue, 28

e-mail: dygalo@vstu.ru

Abstract. The paper describes the approach for the evaluation of predesign-thermal load of the braking mechanism for vehicles with ABS.
Methodology: A method is proposed for determining the energy quenched in a braking mechanism with ABS using three approaches.
Research implications: The study was the basis for creating a computer model of the temperature field of the braking mechanism, which, on the whole, makes it possible to talk about a system for calculating the thermal loading of braking mechanisms with ABS.
Conclusions: Due to the phenomenon of non-synchronous rotation of the car's wheels during braking, due to various factors, the brake mechanisms of even one axle of the car can have different performance indicators along the friction path

The increasing availability of automated braking systems, such as antilock braking systems (ABS), in addition to the obvious advantages in terms of active safety, creates a number of problems caused by changes in the working process. In particular, this applies to the popular in recent years, high-speed cars J – class-crossovers and SUVs, which, as a rule, while maintaining high-speed mode have an increased weight compared to conventional cars. This increases the amount of kinetic energy that must be extinguished during braking.

However, in contrast to the traditional way of stopping a skid, the main part of the kinetic energy of a car with ABS is extinguished by the operation of friction in the brake mechanism, which inevitably leads to an increase in thermal loading, especially when using by the manufacturers of brake systems of cars, the traditional elements of basic models.

At the same time, it is known from literature sources that the overheating of the brake friction pairs contributes to development of a critical feding, accompanied by a sharp decrease (up to 50%) in the coefficient of friction of the brake linings, as well as increased wear of the countertops, with the formation of macroscoles [1]. Therefore, it is necessary to assess the thermal load of friction pairs of brake mechanisms of cars equipped with automated systems.

The analysis of the effect of the principal change in the working process of braking the car wheel with ABS on the distribution of the work necessary to extinguish the kinetic energy of the car as an example is shown in figure 1. The graph shows: the change in the kinetic energy of the car with ABS when braking from an initial speed of 60 km / h on dry asphalt during braking (Wk) , the work spent on friction in contact with the road tire (ATR) and hysteresis losses in the tire (AFF), as well as absorbed in the brake mechanism (ATM).

IOP Conf. Series: Materials Science and Engineering **386** (2018) 012012 doi:10.1088/1757-899X/386/1/012012

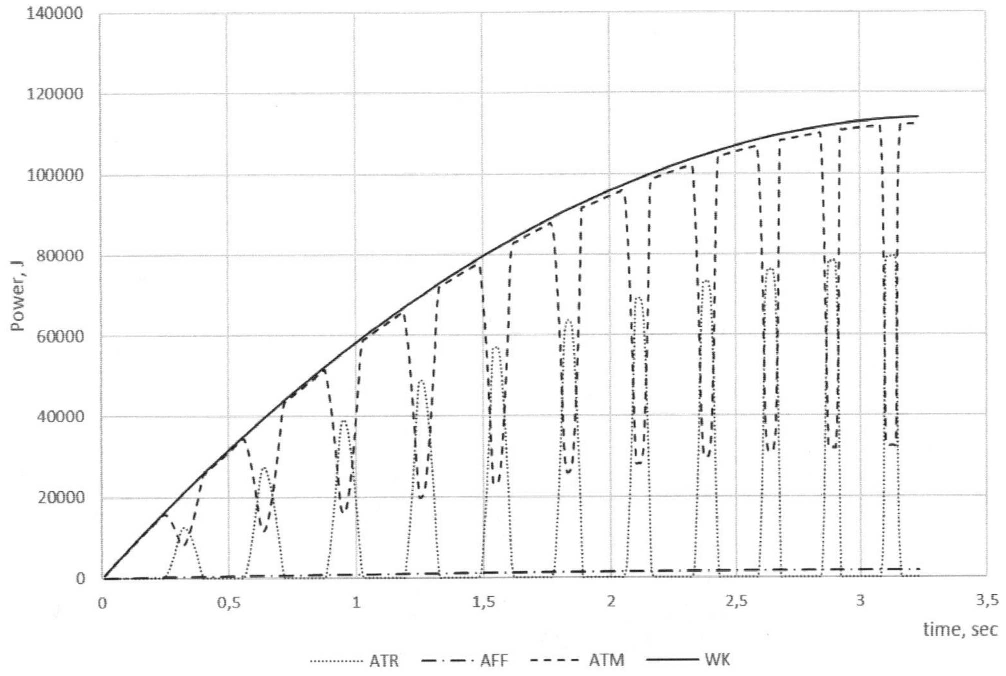

Figure 1. Distribution of work spent on damping the kinetic energy of the vehicle with ABS during braking.

The graph shows that the braking of the car with ABS, in contrast to the skid braking, the main part of the kinetic energy is extinguished by the friction work in the friction pairs of the brake mechanism. Therefore, despite the development of design and calculation technologies, the assessment of the probability of possible failure of the vehicle brakes during operation due to **feding** is an urgent task. It is important that in the design of braking vehicles with ABS manufacturers have the opportunity to assess the effectiveness of their work at the stage of pre-design calculation. The calculation of the average temperature of friction elements in the braking process is of great interest in assessing the temperature of the brake and can be particularly useful in cases where the direct measurement of temperatures at real points of contact and the friction surface as a whole is difficult.

To solve this problem, it is necessary to have computational methods that allow to analyze such performance characteristics as the change in the braking moment of friction, the temperature of the surface of the brake discs, the speed and duration of braking, as well as the work performed depending on the physical, mechanical and thermal properties of the materials of the friction pair, as.

The average temperature of friction pairs, taking into account the short duration of braking, without taking into account heat transfer to the environment, can be determined by the dependence obtained by Professor A.V. Chichinadze [2]

$$\vartheta * = \frac{W_{Т.П.}(1-\alpha''_{Т.П.})b_1}{\lambda_1 A_{a1} t_T}\left[\frac{1}{3}\tau_N + \tau_W F_{01}\right], \qquad (1)$$

where: $W_{Т.П.}$- full braking operation; $\alpha''_{Т.П.}$– the distribution coefficient of heat flow; A_{a1}– the nominal area of friction;

IOP Conf. Series: Materials Science and Engineering **386** (2018) 012012 doi:10.1088/1757-899X/386/1/012012

t_T- duration of braking; $\lambda_{1,2}$- coefficient of thermal conductivity; τ_N, τ_W– time characteristics of power and operation; F_{01}– Fourier number of friction elements.

In turn, the coefficient of distribution of heat fluxes required for the calculation of thermal fields can be found from the expression [2]

$$\alpha''_{T.\Pi.} = 1 / \left[1 + \frac{b_1 \lambda_2}{b_2 \lambda_1} \left(\frac{\frac{1}{3}\tau_N + F_{O2}\tau_W}{\frac{1}{3}\tau_N + F_{O1}\tau_W} \right) \right], \qquad (2)$$

where $b_{1,2}$ is the thickness of friction elements.

The given dependencies allow to calculate the average temperature of brake discs with the help of finite element software [6] with the known full braking operation. Initial parameters in them are parts and Assembly of brake mechanisms, values of heat flows, heat transfer coefficients. Thus, as an example, figure 2 shows the model of temperature distribution of the brake disc of the car with ABS at a coefficient of convective heat transfer of 35 W / m2•K and ambient temperature of 294 K, obtained by using a finite element software..

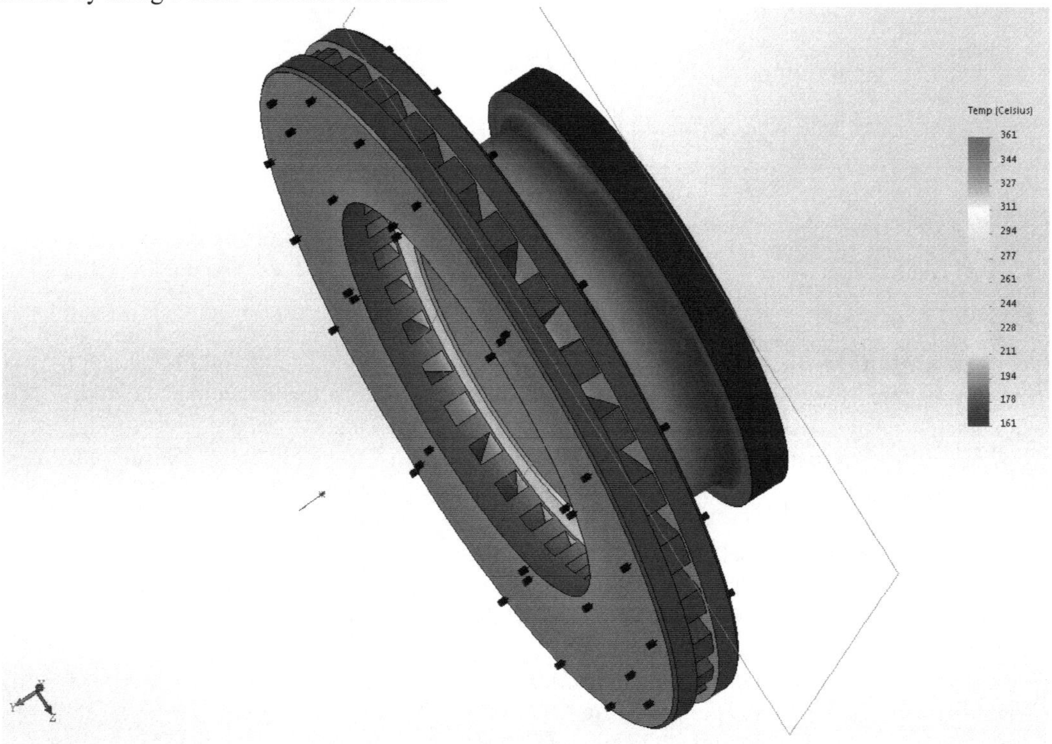

Figure 2. The model of the temperature distribution of the brake disc vehicle with ABS.

As mentioned above, for the application of the dependencies obtained by A.V. Chichinadze [2], it is necessary to find the full braking operation based on the braking force on the disc and the friction path of the brake disc. At the stage of pre-design analysis to determine the braking force, several approaches are possible. The most promising approach is based on the determination of the braking performance brought to the brake disc, which can be found with sufficient accuracy through the ratio of the longitudinal reaction and the braking torque. To do this, we associate them with the value of the dynamic radius. Despite the conventionality of this ratio, this is enough to estimate the amount of work needed. Brake torque on the wheel is determined by the known dependence

IOP Conf. Series: Materials Science and Engineering **386** (2018) 012012 doi:10.1088/1757-899X/386/1/012012

$$M_{m\partial} = R_{x\partial} \cdot r . \quad (3)$$

To determine $\xi_{a\delta c}$ the value of the longitudinal reaction, we will use the degree of using the maximum coefficient of adhesion $\xi_{a\delta c}$ in the operation of the anti-lock braking system [3]. Then the expression for the longitudinal reaction of the wheel will take the form

$$R_x = R_z \cdot \varphi_{\max} \cdot \xi_{a\delta c} . \quad (4)$$

The value is determined from the requirements of standards for the braking performance of the car with ABS. So, for dry asphalt, it should not be below the level of the braking skid $\xi_{a\delta c} = 0,9$, wet and compact snow $\xi_{a\delta c} = 0,85$. Introduction to the calculation of this indicator greatly simplifies the calculation process and eliminates the consideration of the dependencies of the algorithm of a particular anti-lock system, since the algorithm and the matrix of values of the control unit of the anti-lock system are often a trade secret manufacturers of brake systems.

The value of the braking force on the average circumference of the brake disc is found from the expression:

$$P_m = 2\frac{M_{m\partial}}{r_{\partial uc}} \quad (5)$$

where $r_{\partial uc}$ is the radius of the brake disc.

Taking into account the dependence (4), we obtain an expression for the braking force in the form

$$P_m = 2\frac{R_z \cdot \varphi_{\max} \cdot \xi_{a\delta c} \cdot r_{\partial}}{r_{\partial uc}} \quad (6)$$

To find the full operation of friction in the brake mechanism, it is necessary to know the friction path in the "brake disc – pad"pair. Theoretical background to determine the path of friction of individual wheels of the vehicle in braking mode is given in [5].

The rotation of each individual wheel of the vehicle with ABS when braking is characterized by a different combination of speed modes due to the operation of the brake system and the contact of the tire with the road surface, which is shown in Fig. 3.

At the same time, the following characteristic periods can be distinguished: braking, disinhibition, blocking or use of the wheel, free rolling and quasi-uniform movement in the area $\dot{\omega}_{\kappa} \approx 0$.

Each period is characterized by a different combination of speed and load performance. Taking into account the short duration of the ABS cycle, we assume that each period is characterized by a steady slowdown (acceleration) j_{ycm} and an average speed.

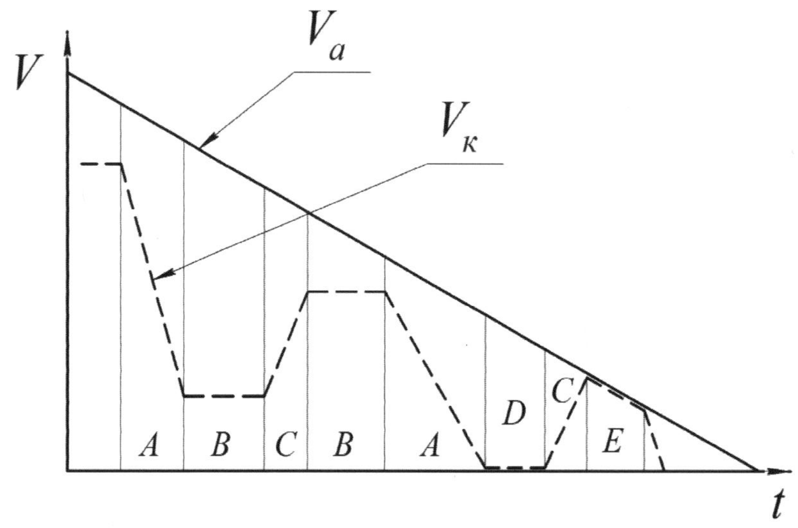

Figure 3. Period options during braking:
A - braking B - quasi-uniform rotation, C - disinhibition, D - Yuz, E - free rotation (full release), $\boldsymbol{V_a}$ – vehicle speed, $\boldsymbol{V_к}$ – given radial rotation speed of the wheel.

With the same values of the braking distance, cars with ABS and without it can be implemented different characteristics for friction pairs "drum (disc)-pad" and "bus-road". To calculate the friction path in each pair of brake mechanisms "disc-pad" use the dependence

$$L_{Tp} = (V_{Н,К} + V_{К,К})(t_{К,К} - t_{Н,К})(r_б / r_К)/2, \quad (7)$$

$$\text{or } L_{Tp} = (V_{Н,К}^2 - V_{К,К}^2)/2j_{уст,к};$$

where: $V_{н,к}$, $V_{к,к}$ - the radial speed of the wheel at the beginning and end of the time interval; $t_{н,к}$, $t_{к,к}$ - the moments of time at the beginning and end of the time interval; $r_б$ - the radius of the brake drum; $r_к$ - the radius of the wheel.

The friction paths for each period are determined from the following dependencies.
The period of inhibition and disinhibition

$$L_{Tp} = (V_{Н,К}^2 - V_{К,К}^2)/2j_{уст,к}, \quad (8)$$

where: $V_{Н,К}$, $V_{К,К}$ - wheel speed at the beginning and at the end of the period, $j_{уст,к}$ - the steady-state acceleration of the wheel, $r_к$ - wheel radius, $r_б$ - the average radius of the brake disc.
Free wheel rolling $L_{Tp} = 0$,

Wheel lock $L_{Tp} = 0$

Constant (quasi-uniform) rotation of the wheel

$$L_{Tp} = V_{Н.К} \cdot t_П \quad (9)$$

where: $t_П$ - the duration of the period of uniform rotation of the wheel.
To find the General friction path in all areas, let's sum the above dependencies

$$L_{Tp,тм} = \sum_{i=0}^{I} L_{TpTi} + \sum_{i=0}^{J} L_{TpPj} + \sum_{i=0}^{N} L_{TpПn}, \quad (10)$$

where: $L_{mp,\ m\text{м}}$ - the friction path of the lining of the drum; I - the number of braking areas; J – the number of areas of disinhibition; N – the number of areas of constant rolling of the wheel;

L_{TpTi} – the friction path for a pair of "disc-pad" on the braking area;

$L_{Tp\,Pj}$ - way friction pair "disc pad" on the j-th section of the release;

$L_{Tp\,\Pi n}$ - the friction path for the pair "disc-pad" on the n-th section of the wheel constant rolling braking.

Finally, the friction work brought to the brake disc is determined from the expression:

$$W_{T.\Pi.} = L_{Tp} P_m \,, \qquad (11)$$

Due to the phenomenon of non-synchronous rotation of the wheels of the car during braking, due to various factors, both in the presence of ABS and without it, the brake mechanisms of even one axis of the car can have different indicators of work on the friction path.

It should be noted that the definition of the full braking performance and, as a consequence, the determination of the energy quenched in the brake mechanism was the basis for the creation of a computer model of the temperature field of the brake mechanism, which generally makes it possible to talk about the system for calculating the thermal loading of the brake mechanisms with ABS.

References

[1] Bezyazychnyy V.F.. Lyubimov R.V.. Timofeyev. 2000. Eksperimentalnoye izucheniye protsessov razrusheniya poverkhnostnykh sloyev metalla pri ustanovivshemsya protsesse fretting-iznashivaniya. *Sb. nauchn. trudov Tverskogo gos. tekhn. un-ta «Mekhanika i fizika friktsionnykh kontaktov».* vyp. **7**. p.24-28.

[2] Chichinadze A.V. 1970.Teplovaya dinamika treniya. 171 p.

[3] Revin A.A., Dygalo V.G., 2014. Formirovaniye osnovnykh ekspluatatsionnykh svoystv avtotransportnykh sredstv v rezhime tormozheniya. *Avtomobilnaya promyshlennost.* **11**. p. 3-5.

[4] Revin A.A., Poluektov M.V., Radchenko M.G., Zabolotnyy R.V. 2013. Vliyaniye rabochego protsessa ABS na dolgovechnost elementov shassi avtomobilya: monografiya. 224 p.

[5] Tumasov A.V., Groshev A.M., Kostin S.Yu., Saunin M.I., Trusov Yu.P., Dygalo V.G., 2011. Issledovaniye svoystv aktivnoy bezopasnosti transportnykh sredstv metodom imitatsionnogo modelirovaniya. *Zhurnal avtomobilnykh inzhenerov.* **2**. p. 34-37.

[6] Turbin I.V., Epishkin V.E., Solomatin N.S. 2014. Vliyaniye koeffitsiyenta treniya na tribotekhnicheskiye kharakteristiki par treniya diskovogo tormoza. *Materialy konferentsii: «Perspektivnyye napravleniya razvitiya avtotransportnogo kompleksa : sbornik statey VIII Mezhdunarodnoy nauchno-proizvodstvennoy konferentsii.»* p. 124-128.

[7] S. Voloaca. Concerns regarding temperature distribution obtained by experiments and finite element analyses for types of brake discs/ S. Voloaca, G. Fratila/ U.P.B. Sci, Bull., Series D, Vol. 74, Iss. 3, 2012. – p. 33

IOP Conf. Series: Materials Science and Engineering **386** (2018) 012013 doi:10.1088/1757-899X/386/1/012013

Geoinformation Support of Ground Vehicles' Autonomous Driving

S S Shadrin

D.Sc. (Engineering), Head of R&D Dept., Basetrack Technologies Limited, Cyprus

E-mail: dr.shadrin@basetrack.net

Abstract. This article describes relevant challenges of autonomous driving with respect to the strategic and tactical levels of vehicle control, some solutions are offered. Methods for creating a digital terrain model are described. The notion of a "basetrack" is introduced, which is a high-accuracy spatial driving route that contains additional data sets for the vehicle's optimal control with respect to the requested transport task, road and traffic conditions. A "basetrack" data structure is presented with respect to the advantages of its implementation: the predetermined variability of control actions on the route, depending on the tasks of time saving, energy efficiency, or safety; increased reliability, reduced requirements for on-board computing power, and improved performance. We propose to solve the issues in optimal vehicle control for the autonomous driving mode in simulation with reference to a precise geoinformation environment.

1. Introduction

Energy efficiency and safety are traditionally among the main development trends in the automobile industry. The opportunities for automakers to improve the principal design and control algorithms seem to be exhausted, and it is a challenge to improve the safety of road vehicles. The analysis of road accidents involving heavy trucks showed that 85.2% of road accidents are caused by various human factors in France, Germany, Italy, Hungary, the Netherlands, Slovenia, and Spain [1]. Approximately 80% of road accidents in Russia are caused by driver's faults. Thus, the most efficient approach could be increasing the active safety of road vehicles and excluding human drivers from the cycle of vehicle control.

Development projects for autonomous vehicle technologies are currently being undertaken by all global car manufacturers, especially in the US, Germany, Japan, and China. Automobile groups are involved into these projects as well, such as Ford, Daimler, Volkswagen, Toyota, Honda, GM, Geely, Tata, and Tesla, as well as other major technological organizations such as Google, Continental, Delphi, Siemens, and Bosch. Other players in the field include defense departments and agencies of different countries (e.g., DARPA [2]), universities (e.g., Stanford University, Carnegie Mellon University, Technical University of Munich, University of Karlsruhe, Fraunhofer Institute, University of Minnesota, Universidad Politécnica de Madrid [3]), and many other institutions [4-6]. Thus, the development trends of autonomous driving technologies are important and relevant.

2. Problem Statement

The overall act of driving can be divided into three levels of a driver`s effort [7, 8]: strategic, tactical, and operational levels. The strategic level involves trip planning. The tactical level involves maneuvering the vehicle through traffic (lane changes, overtaking, speed limit, etc.). The operational level involves innate actions such as steering, braking, and other vehicle system control for trajectory tracking.

The strategic level deals with tasks of logistics, routing, navigation [2, 5, 6, 9]. One of the main applied problems at this level relates to the creation of high-accuracy models of geoinformation about the spatial terrain, which are called high-definition (HD) maps.

The tactical level deals with the issues of psychophysiology and critically depends on the perfect functioning of the technical system of road scene recognition [10], including the subsystem of technical vision [11]. In addition, there is a huge layer of decision-making problems at the tactical level, which researchers and developers mostly try to solve using artificial intelligence systems [12]. The main problems of the tactical level include all the uncertainties that arise during the process of driving, which require route correction. The operational level of vehicle control is supposed to be well studied [2, 3, 5, 6, 10, 13].

This study focuses on new approaches that solve some challenges of tactical and operational levels in autonomous vehicle control.

3. Geoinformation Support as an Instrument for Rapid Implementation and Wide Spread of Autonomous Vehicles

The known methods of autonomous driving are mostly based on the use of data from various real-time navigation systems and an electronic map loaded into the ECU memory. The exact driving control actions are commonly carried out using computational algorithms that operate in real time and often depend on artificial intelligence systems. These systems require significant on-board computing power for tasks such as object detection, classification, scene recognition, risk assessment, and decision making. Trajectory correction in such systems continually occurs on each segment of the route, which results in high on-board computational requirements, increased lag for control commands with the increase of driving velocity, decreased control accuracy at high velocities, and a risk of unpredictable failures in cases of wrong scene recognition.

We offer a simplified approach for autonomous driving and a more reliable and less expensive solution. We propose the concept of a "basetrack," which is a high-precision spatial driving route that contains additional blocks that provide optimal control of the vehicle based on the requested transport task and traffic conditions. The basetrack can be compared with a virtual spatial railway for the vehicle, which acts like a train. Accordingly, in the simplest case, the vehicle should decrease its driving velocity and stop in the event of an obstacle or risk detection. If you have trouble, the decision-making system (e.g., the Mivar expert system [12]) should start to switch the trajectory tracking system to another track from the database or to another fragment of a generated detour track.

There are different approaches to HD-map design, but all of them begin with the creation of a digital terrain model (DTM). Informational layers can be added to the DTM with corresponding objects such as road markings and infrastructure elements. Geospatial data for DTM can be obtained by space or aerial photography, laser scanning (ground or air based), radar scanning, and other methods. However, we consider the use of traditional HD maps to be unreasonable in the proposed conceptual approach. Instead thatwe propose the use of basetracks, which are software products designed based on DTM and technical knowledge about a certain vehicle, road, traffic, and environmental conditions.

Figure 1 illustrates the principal scheme for autonomous vehicle control with the use of a basetrack.

IOP Conf. Series: Materials Science and Engineering **386** (2018) 012013 doi:10.1088/1757-899X/386/1/012013

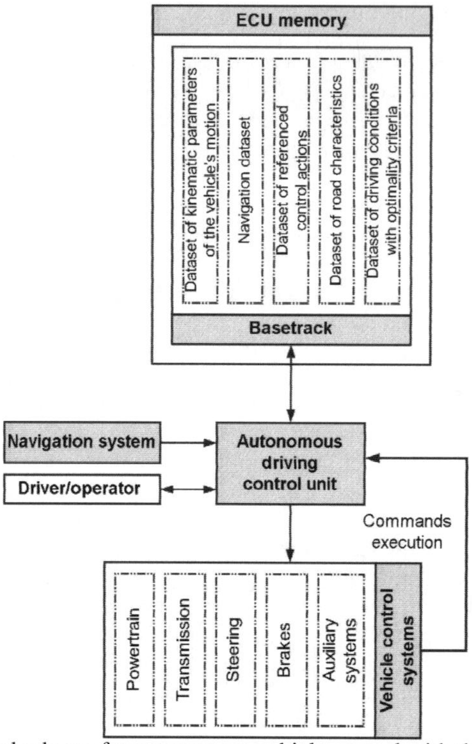

Figure 1. Principal scheme for autonomous vehicle control with the use of basetrack.

The basetrack structure includes:

- a dataset of kinematic parameters of the vehicle motion;
- a navigation dataset;
- a dataset of referenced control actions;
- a dataset of road characteristics;
- a dataset of driving conditions with optimality criteria.

The dataset of kinematic parameters of the vehicle motion contains calculated referenced data that are typical for the vehicle considered when driving in normal conditions (no extra wind load, loaded car weight, dry asphalt, etc.).

The navigation dataset enables vehicle positioning on the basetrack while driving with the help of any precise on-board navigation system.

The dataset of referenced control actions corresponds with the dataset of kinematic parameters and contains control commands for normal driving conditions.

The dataset of road characteristics contains information on the road type, longitudinal and lateral inclinations, average road surface friction coefficient in normal weather conditions, International Roughness Index (IRI), and road curvature.

The dataset of driving conditions contains information that is included but is not limited by the following:

- the maximum permissible speed V_{max} in accordance with traffic rules and road signs;
- the coefficient of weather conditions K_w, by which the permissible speed can be adjusted;
- the critical speed V_{crit} for a certain vehicle (the maximum possible speed for the vehicle and road conditions);
- the complex rate K_{eff} of speed adjustment, which reflects the criterion of energy efficiency;
- the coefficient K_{infr} of speed adjustment, which takes into account the peculiarities of road infrastructure;
- the coefficient K_{sat} of the reliability degree for received satellite navigation data at a given location. This factor takes into account the infrastructural and natural disturbances and is an indicator

of switching from a satellite system to another positioning system (optical, inertial, etc.) when using a hybrid navigation system.

The dataset of driving conditions can be updated interactively.

Importantly, some parameters of the driving condition dataset are mutually exclusive. Accordingly, specific control actions for autonomous driving mode will be determined from the considered dataset with respect to the transport task. Some examples are presented by optimality criteria such as the minimum travel time, fuel economy, or safety driving with risks reduction. In the case of a race car driving on a circuit, the autonomous driving control unit will use the speed determined as $V_{crit} \cdot K_w$ from the dataset of driving conditions. In another example, a heavy truck driving autonomously with the task of minimizing fuel consumption will cause the control unit to use speed determined as $V_{max} \cdot K_{eff} \cdot K_w$.

A set of files with sequential basetracks forms a track of a larger length. Before the start of autonomous driving, the track file or at least the first file of a long track should be loaded into the on-board memory block. The rest of the track files and the dataset of driving conditions can be downloaded or updated as the vehicle drives along the route. It is also possible to preload all track files into the memory block.

We propose using a basetrack concept at the strategic level of autonomous vehicle control while considering the information of modern HD maps to be redundant. The use of basetracks at the tactical level excludes or minimizes the problem of decision-making and dynamic route correction, which leads to increased reliability of the autonomous driving system, reduces requirements for on-board computing power, and improves the system performance, allowing it to drive the vehicle at higher velocities.

For the prototype of an autonomous road vehicle, speed of 130 km/h was achieved for stable driving in the corridor of road markings using the basetrack in autonomous mode. This was achieved without the use of road marking recognition systems in accordance with the experiment's goal, which was to demonstrate the pure basetrack technology on a real road.

At the same time, the basetrack concept provides different predetermined ways of driving along the same route. An economic advantage is obtained based on the correlation of optimized control actions in autonomous mode with respect to the requested transport task.

4. Conclusion

The results of this study are as follows:

1. The notion of a basetrack has been introduced, which is a high-accuracy spatial route of driving that contains additional datasets for optimal vehicle control in autonomous mode with respect to the requested transport task, road conditions, or traffic conditions.
2. We proposed using basetracks at the strategic and tactical levels of autonomous vehicle control instead of HD maps or as a separate layer of an HD map.
3. The use of a basetrack concept at the tactical level excludes or minimizes the problem of decision-making and dynamic route correction, which could increase the reliability of autonomous driving systems, reduce requirements for on-board computing power, and improve the system performance, allowing vehicles to operate at higher velocities.
4. It is essential to solve the issues in optimal vehicle control for the autonomous driving mode in simulation with reference to a precise geoinformation environment.

References

[1] Shadrin S.S., Ivanov A.M., Karpukhin K.E. Radical improvement of road safety by integrating autonomous wheeled vehicles into an intelligent transport environment. Bulletin of Machine Building, 2018, No. 1: 85-88.

[2] Buehler M., Iagnemma K., Singh S. "The DARPA Urban Challenge. Autonomous Vehicles in City Traffic," Springer, 2009: 651 p.

[3] J Naranjo. E., Clavijo M., Jiménez F., Gómez O., Rivera J. L. and Anguita M. "Autonomous vehicle for surveillance missions in off-road environment". IEEE Intelligent Vehicles Symposium (IV), Gothenburg, 2016: 98-103.

[4] Saikin A. M. et al. "Tendency of creation of 'driverless' vehicles abroad," Biosciences Biotechnology Research Asia, 2014, vol. 11 (Spl. Edn.): 241–246.

[5] "Autonomous Control Systems and Vehicles – Intelligent Unmanned Systems" K. Nonami et al.

IOP Conf. Series: Materials Science and Engineering **386** (2018) 012013 doi:10.1088/1757-899X/386/1/012013

(Eds.), Springer Japan, 2013: 306 p.
[6] Cheng H. "Autonomous Intelligent Vehicles – Theory, Algorithms, and Implementation," Springer, 2011: 163 p.
[7] Michon J. A. "A critical view of driver behavior models: What do we know, what should we do?" In L. Evans & R. C. Schwing (Eds.), "Human Behavior and Traffic Safety," New York: Plenum Press, 1985: 485–520.
[8] SAE J3016: Taxonomy and Definitions for Terms Related to Driving Automation Systems for On-Road Motor Vehicles, U.S.A.: SAE, 2016: 30 p.
[9] Tahirovic A., Magnani G. Passivity-Based Model Predictive Control for Mobile Vehicle Motion Planning. Springer, 2013: 64 p.
[10] Shadrin S.S. Methodology of autonomous road vehicles, integrated in intellectual transport environment, driving control systems design: dissertation ... doctor of technical science. 05.05.03 [Place of the thesis defense: Bauman State Technical University]. Moscow, 2017: 400 p.
[11] Saikin A.M., Buznikov S.E., Karpukhin K.E. "The Analysis of Technical Vision Problems Typical for Driverless Vehicles," Research Journal of Pharmaceutical, Biological and Chemical Sciences, 2016. Vol. 7, No. 4: 2053-2059.
[12] Shadrin S.S., Varlamov O.O., and Ivanov A.M. "Experimental Autonomous Road Vehicle with Logical Artificial Intelligence," Journal of Advanced Transportation, 2017, Article ID 2492765, 10 pages, doi:10.1155/2017/2492765.
[13] Shadrin S.S., and Ivanov A.M. "Algorithm of Autonomous Vehicle Steering System Control Law Estimation while the Desired Trajectory Driving," ARPN Journal of Engineering and Applied Sciences, 2016, Vol. 11, No. 15: 9312-9316.

IOP Conf. Series: Materials Science and Engineering **386** (2018) 012014 doi:10.1088/1757-899X/386/1/012014

Impact of diesel fuel and palm oil blend compositions on the performance of the fuel supply process in the diesel engine

E A Salykin[1], A V Kurapin[1], K E Tshibanda[3], V G Dygalo[2] and V M Slavutskiy[2]

[1] PhD, Associate professor, Volgograd State Technical University, Russia, 400005, Volgograd, Lenin avenue, 28
[2]D. Sc., Professor, Volgograd State Technical University, Russia, 400005, Volgograd, Lenin avenue, 28
[3]Graduate student, Volgograd State Technical University, Russia, 400005, Volgograd, Lenin avenue, 28

E-mail: esalykin@gmail.com, dygalo@vstu.ru

Abstract. The world experience of using palm oil as a fuel for diesel engines was considered. The influence of the mixture composition on the indices of the injection process at a constant value of the active stroke of the plunger and the mass cyclic feeding of the mixture, corresponding to the nominal cyclic feeding in the case of using only diesel fuel is analyzed. A comparative analysis of the change in the parameters of spraying of diesel fuel mixtures and palm oil fractions from 0% to 60% is performed depending on the specific parameters of the fuel supply equipment and the mixture (working stroke of plunger, cyclic feed and heat input). It is shown that changes in the fuel supply equipment parameters when using mixed fuels based on palm oil are not required, since the corresponding change in the volume cyclic feed and active stroke of the plunger with a change in the proportion of palm oil in the mixture to 60% does not exceed 3%. According to the requirements of the standard, these deviations are within acceptable limits.

1. Introduction

In the last years, the leading countries of the world are giving a great deal of attention to the possibilities in the application of internal combustion engine biofuels made from vegetable oil feedstock in order to reduce the dependence of petroleum resources and significantly reduce greenhouse gas emissions [2].

Using biofuels in diesel engines is called biodiesel fuel or biodiesel. Biodiesel fuel can be used as a mixture with diesel fuel of petroleum origin, and in pure form. Each country has various kinds and stocks of biofuels and their use largely depends on the energy program of this country.

The most common types of biofuels are vegetables oils, therefore they are studied as diesel fuel substitutes and energy sources all over the world. In the European Union (EU) and the United States biodiesel fuel is called oil methyl ester from vegetable oils, basically rapeseed oil. The methyl ester of rapeseed oil as biodiesel is widely used in France, Austria and Germany [1, 3, 4, 19, 20, 21]. Esters of soybean oil are used in the US.

An actively developing research on the application of vegetable oils as a promising source of energy for diesel engines is also being carried out in Russia [1, 3, 4].

In the countries with tropical climate production of palm oil is widespread.

In Malaysia, Indonesia, Thailand, Colombia, Nigeria and other countries, where the conditions for growing oil palms are perfect, interest in the oil palm is increasing. In these countries the technical aspects of palm oil as an alternative fuel for the diesel engine are extensively studied and research is going on [14, 15, 16, 17]. A large number of research papers on engine performance and emission characteristics using a wide range of biodiesel feedstocks have appeared in the last decades . In this paper some results of the research carried out are observed.

1.1. Literature review

Mosarof M H and others [14] reported that many investigations around the world were carried out to study economic aspects, performance emission, wear characteristics and the effects of fuel properties such as cetane number, aromatic content and viscosity on emission characteristics. According to Ziejewski and Kaufman [12], most researchers have concluded that vegetable oils used as motor fuel, can burn safely during short-time combustion in diesel engine. Others authors showed that high cetane number in a biodiesel methanolysis of ether or palm oil (51 and more) contributes to the reduction of the period of time delay of ignition and lessens the "hardness" of the engine's operation. Vedruchenko V R et al [6] carried out research based on analysis of effects of fundamental physics properties for same selected kinds of diesel fuels on process injection in the higher pression systems. They found that, the properties of using fuel in operating diesel engine have the effects on different indicators of power and efficiency of the engine. It is necessary to bring the stabilization into the process of injection and ignition using an effective device for controlling the viscosity, density and the compressibility.

Markov V A and others [7] carried out the experimental study, the analysis of which shows that, the dynamic development of the fuel jet for an operating diesel engine fuelled by mixed fuel has dependence on the concentration of components of a fuel blends. The maximum length of the jet, the maximum width and the opening angle of the jet during the period of ignition delay depends on the volumetric contents of components in the mixture of diesel and vegetable oils.

Yusaf and others [17] investigated the performance and emission of a 4-cylinder perking diesel engine fueled with 25%, 50% and 75% crude palm oil (CPO) blends. At higher engine speed, they found lower brake-specific fuel consumption (BSFC) and higher torque for the CPO blends, but the brake power produced was slightly lower. Oxygen and nitrogen oxides (NO_x) emissions were reduced, while carbon mono-oxide (CO) and exhaust gas temperature (EGT) were increased for CPO blends compared with diesel fuel. Ndayishimiye and Tazerout [11] in 2011 conducted a study of the performance and assessment of the exhaust gases of single-cylinder diesel engine with direct injection combustion chamber, mixtures of crude palm oil and heated palm oil and mixtures of palm biodiesel and diesel fuel. Slightly higher values of specific effective fuel consumption for all tested biofuels in comparison with pure diesel fuel were detected. Effective efficiency of the engine is somewhat increased when using mixtures of crude palm oil, but decreases in all other tested biofuels. Emissions of CO and hydrocarbon (HC) have been significantly reduced (by 30...65%) for palm biodiesel, while NO_x emissions are increased. Rashid et al [18] in 2015 explored the work of an operating diesel engine fueled by biofuels based on palm (PB20), jatropha (JB20) and methyl ester moringa oil (MB20), in comparison with diesel fuel (B0). The results showed that the decrease in the effective power was up to 8.03 % for MB20, up 6.92 % for PB20, of 8.75 % for JB20 compared to B0. The increase in the effective specific fuel consumption was 8.39 % for MB20, of 7.15 % for JB20 and of 5.42 % for PB20. The reason for this is the low calorific value of the oils**.** Lowering CO emissions are 22.93 % – MB20, 27.23 % – JB20 and 32.65 % – PB20. This is due to the presence of a larger percentage of oxygen in the oil. Lower HC emissions are 30.26 % – PB20, 19.73 % of JB20, 11.84% – MB20**.** It is also due to higher oxygen content, high cetane numbers and high in saturated fatty acids. The increase in NO_x emissions is 6.91 % – PB20, 14.22 % – JB20 and 18.56 % – MB20. This is due to earlier combustion of biodiesel fuels and hence higher temperature at the end of combustion due to higher cetane numbers and higher content of oxygen in them. The results also showed the advantage of using a mixture with palm oil in all indicators.

IOP Conf. Series: Materials Science and Engineering **386** (2018) 012014 doi:10.1088/1757-899X/386/1/012014

Globally, a large number of literature from highly rated journals in scientific indexes were reviewed, including the most recent publications on palm oils as a biodiesel [14, 15, 16, 18].

However, it was observed that numerous investigations in palm oil as a biodiesel fuel substitute have presented the tribological, environmental and economic aspects of using this type of oil. The effects of the palm oil, properties of its mixture with diesel fuel influencing the performance of fuel injection process in the fuel supply system have not been exhaustively analyzed in special literature, some efforts were made in only few papers.

1.2. The purpose and objectives

The paper presents numerical studies of effect of composition properties of fuel mixture on operation of fuel injection process in the fuel supply system, using selected technical solutions for the stabilizations of fuel supply process and the calorific value of the mixture, and the following problems have been solved:

1 – analysis of the effect of physical properties of palm oil in a mixture with diesel fuel mark L;

2 – analysis of the impact of structural-adjustment parameters of the high pressure pump when operating on mixed fuel. Using diesel fuel mark L active stroke of the plunger h_{act} and injection quantity q_{cycl} have been recorded. The values of these parameters correspond to those engines operating on diesel fuel.

3 – the analysis of the processes of fuel injection, while maintaining constant the calorific value of the mixture Q_{mix} = const, (H_{mix} = var) variables q_{cycl} and h_{act}.

2. Object and methods of research

In this research diesel engine type D-144 made in Vladimir tractor factory (VTZ, Russia) as an object of numerical studies is considered.

It is equipped with fuel pump type UTN-5 with the diameter of the plunger d_{pl} = 8 mm, plunger stroke h_{pl} = 8 mm, pump rotation speed n_{nom} = 1000 mn^{-1} , injectors type FD-22 with the standard nozzle and the pressure of the beginning injection is 17 MPa. The engine specifications are presented in table 1.

Table 1. Technical specification of the test engine

Working	compression ignition, 4 stroke
Diesel type	direct injection, air refrozen
Number and arrangement of cylinder	4, inline vertical
Bore and Stroke	105×120 mm
Displacement	4,15 l
Compression ratio	16,5
Maximum net power	44,1 kW at 2000 rpm
Maximum net torque	221,4 N·m at 1550 rpm

As a base fuel diesel fuel (D100) is used and palm oil (PO) is used as addition to the base fuel. Mixture proportion is 10, 20, 30, 40, 50, 60 vol % of palm oil addition. We denote these mixtures, respectively PO10, PO20, PO30, PO40, PO50, PO60. Pure palm oil we denote as PO100. The basic properties of fuel blends are shown in table 2.

In order to assess the impacts of mixed fuel properties in the fuel supply system during transition from diesel fuel to mixed one, a numerical experiment was carried out using mathematical model integrated in the program complex "PC-injection", developed at the Department "Diesel engines" Moscow State Technical University Bauman.

The mathematical model underlying the PC is based on tested approaches and results of original fundamental research in modern injection modeling machines.

"PC Injection" taken into account here also applies the law of fuel supply and reflects the impact of fuel properties and the blends compositions on the dynamic indicators of the process of fuel injection in fuel supply system [9].

Table 2. Properties of diesel fuel, palm oil and their blends

Properties	D100	PO100	PO10	PO20	PO30	PO40	PO50	PO60
Kinematic viscosity at 40 °C (cSt) at 100 °C (cSt)	3.9	- 8.6	4.37	4.84	5.31	5.78	6.25	6.72
Density at 20°C (g/cm^3)	830.0	918.0	838.8	847.6	856.4	865.0	874.0	882.8
Flash point (°C)	250	315	260	270	280	290	300	289
Calorific value (MJ/kg)	42.5	37.1	42.01	41.33	40.73	40.32	39.98	39.20
Theoretical quantity of air required for ignition 1kg of fuel. (kg/kg)	14.45	12.7	14.14	13.98	13.82	13.66	13.5	13.34
Oxygen content (% by mass)	0.4	11.4	1.5	2.6	3.7	4.8	5.9	7
Hydrogen content (% by mass)	12.6	12.0	12.54	12.48	12.42	12.36	12.3	12.24
Carbon content (% by mass)	87	76.6	85.26	84.8	83.88	82.84	81.8	80.76
Cetane number	46.5	49.0	46.75	47.0	47.25	47.5	47.75	48.50
Heat emitted fuel (J)	2539.8	2452.1	2534.7	2520.5	2501.2	2511.1	2515.8	2489.3

For the numerical experiment, the following parameters have been used: nominal volumetric cyclic flow 72 mm^3, the cyclic mass flow of 0.06 g and density of diesel fuel 830 kg/m^3. The values of the parameters correspond at nominal parameters of the engine operation with the diesel fuel.

To identify several factors affecting the efficient performance of the fuel supply, numerical experiments were carried out to study the multifactorial designs tests.

In the first stage, a series of numerical experiments aimed at investigating the effect of composition of diesel fuel mixture with palm on the characteristics of the fuel injection during fixation of the active stroke of the plunger were performed h_{act} = const.

In the second phase, the calculation was made to assess of the impact of composition blend on the characteristics of fuel injection process when fixing cyclic flow (q_{cycl} = const). A constant value of the active plunger stroke (h_{act} = const) and cyclic supply flow of fuel (q_{cycl} = const) corresponds to those in the engine operating on diesel fuel.

At the third stage of the numerical experiment we studied the influence of fuel mixture on the basis of diesel fuel with proportion of 10 to 60 vol % of the palm oil to the fuel injection parameters, while maintaining the calorific value of the mixture $Q_{cм}$ = const, (H_u = var) variables q_{cycl} and h_{act}.

3. Results and discussion

In general, it should be noted that the increased density and viscosity of palm oil are the reason for the increase in mass of the cyclic flow, and the time flow of mixed fuel when it is used in a diesel engine in comparison with diesel fuel according to standards GOST 305-2013. This is confirmed by the following calculations. Calculate the density (ρ_b) and mass cyclic flow ($G_{cycl.b}$) for mixed fuel consisting of diesel fuel and palm oil. Take the density of diesel fuel 830 kg/m^3, density of palm oil is 918 kg/m^3 under normal conditions, volumetric cyclic fuel consummation – 72 mm^3. The results of the calculations are shown in Fig. 1a.

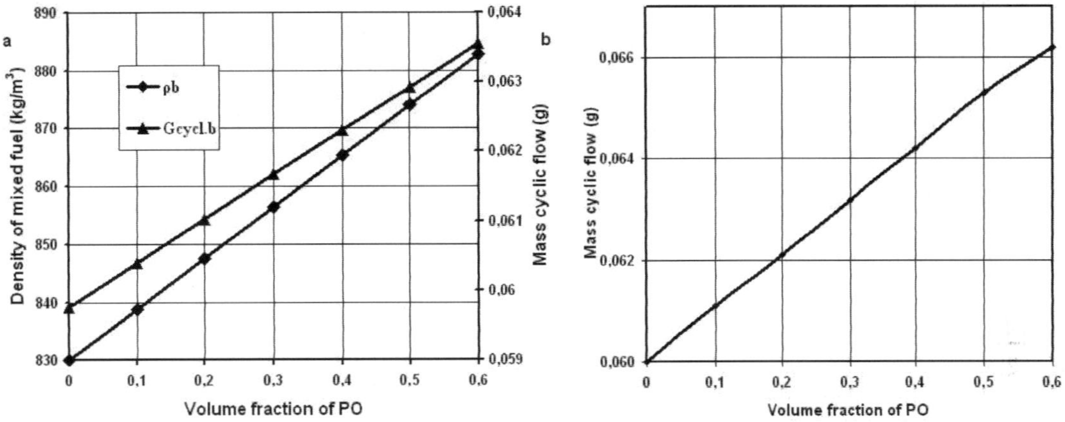

Figure 1. (a) Dependence of the density ρ_b, and mass cyclic flow $G_{cycl.b}$ on volume fraction PO g_{po}; (b) Dependence of the mass cyclic flow $G_{cycl.b}$ on volume fraction PO g_{po} at a constant active stroke of the plunger

In the first time, it was performed the calculations of the parameters of the fuel supply process for the mixed fuel at a constant active stroke of the plunger corresponding to the nominal cyclic feeding on pure diesel fuel, which is equal to 1.93 mm. The results of the calculations are presented in Fig. 1b and 2a. The analysis of these figures shows, that the increase of volume proportion of palm oil in the composition of the mixed fuel leads to the increase in mass of the cyclic flow of the mixture, the increase of average and maximum injection pressures and the total delivery time. This is due to the increase in the density and viscosity of the mixture. It should be noted that a greater increase in mass of the cyclic flow $G_{cycl.b}$ of the mixture with the increase volume proportion of palm oil in the mixture was obtained in the hydrodynamic calculation (see Fig. 1b), in comparison with the mass of the cyclic flow $G_{cycl.b}$ of the mixture obtained only by taking into account the density of the mixture (see Fig. 1a). This can be explained by a reduction in the leakage through the gaps in the piston pair due to the increase in viscosity of the mixture and the growth of palm oil volume proportion g_{po}.

For the reason of e stabilization of the cyclic mass flow $G_{cycl.b}$ of the mixture fuel with the increase of the volume proportion of palm oil in the mixture g_{po}, it is necessary to reduce an active stroke of the plunger, that is, to adjust the fuel equipment (see Fig. 3a). On that basis, a series of calculations of indicators of processed fuel were performed for the mixed fuel under the condition of constant mass of the cyclic flow $G_{cycl.b}$ of the mixture with the corresponding cyclic feeding on pure diesel fuel, which is equal to 0.06 g. The results of the calculations are shown in Fig. 2b.

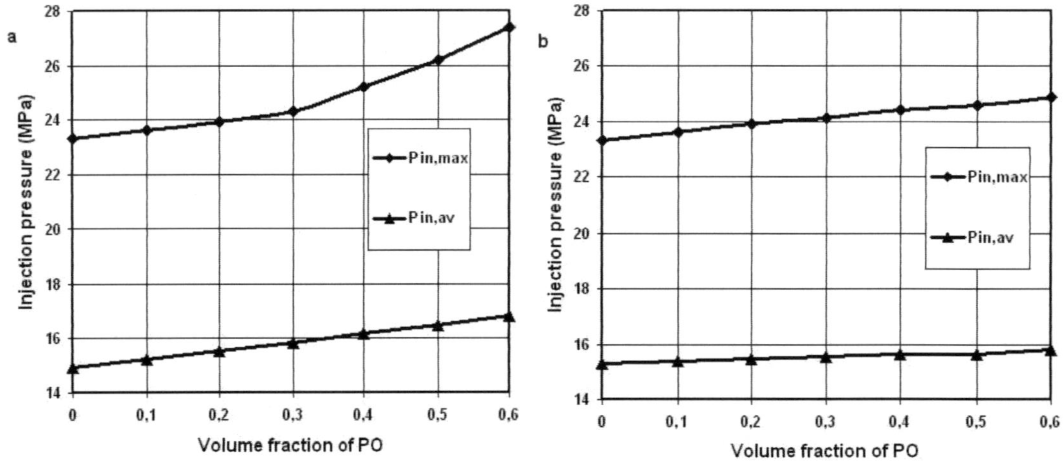

Figure 2. (a) Dependence of the average $P_{in.av}$ and maximum $P_{in.max}$ injection pressureson volume fraction PO g_{po} at a constant active stroke of the plunger; (b) Dependence of the average $P_{in.av}$ and maximum $P_{in.max}$ injection pressureson volume fraction PO g_{po} at a constant mass cyclic flow

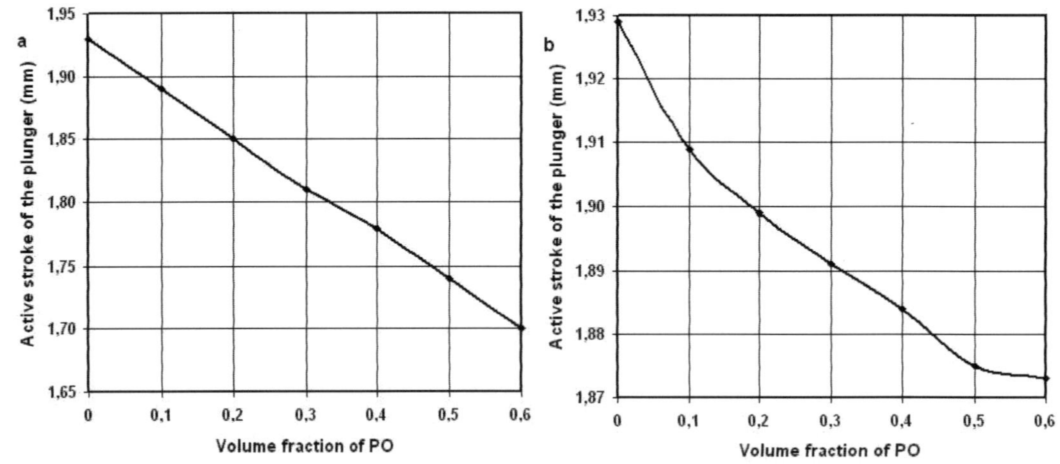

Figure 3. (a) Dependence of the active stroke of the plunger on volume fraction PO g_{po} at a constant mass cyclic flow; (b) Dependence of the active stroke of the plunger on volume fraction PO g_{po} at a constant amount of introduced heat

As can be seen from Fig. 3a, the active stroke dropped from 1.93 to 1.7 mm, however, the maximum and average injection pressures continue to increase with the increase of the volume proportion of palm oil in the mixture g_{po} up to 24.9 MPa and 15.8 MPa respectively (see Fig. 2b), due to the increase of density and viscosity of the mixture.

Stabilization of the cyclic mass flow $G_{cycl.b}$ of the mixture with the increase of the volume proportion g_{po} of palm oil in the mixture leads to a decrease in the amount of heat generated by the fuel in the combustion chamber along with a reduction of the calorific value of the mixture, which leads to lower power diesel. Figure 6 shows the change of in the combustion chamber at a constant cyclic mass flow $G_{cycl.b}$ of the mixture, which is equal to 0.06 g. The calorific value of diesel fuel is taken equal to 42500 kJ/kg, palm oil – 37100 kJ/kg.

In this regard, it seems appropriate to adjust the active stroke of the plunger with the constant amount of heat generated by the fuel into the combustion chamber. We have performed calculations of the parameters of the fuel supply process for the mixed fuel in consideration with these of these circumstances for a constant amount of the heat Q_b, equal to 2.54 kJ, and corresponding to pure diesel fuel at a nominal cyclic flow and is. The results of the calculations are presented in Figures 4b, 3b and 5.

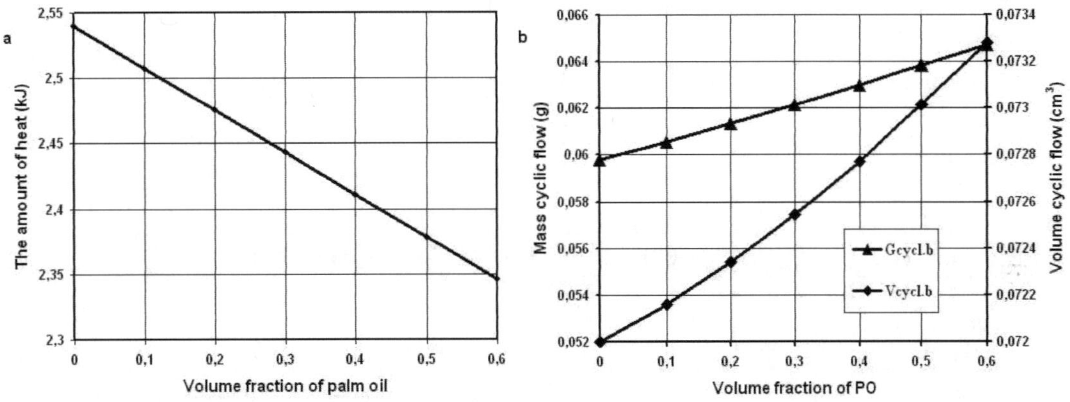

Figure 4. (a) Dependence of heat generated by the fuel in the combustion chamber on volume fraction PO g_{po} at a constant cyclic mass flow of the mixture; (b) Dependence of mass cyclic flow $G_{cycl.b}$ and volume cyclic flow $V_{cycl.b}$ on volume fraction PO g_{po} at a constant amount of the heat generated

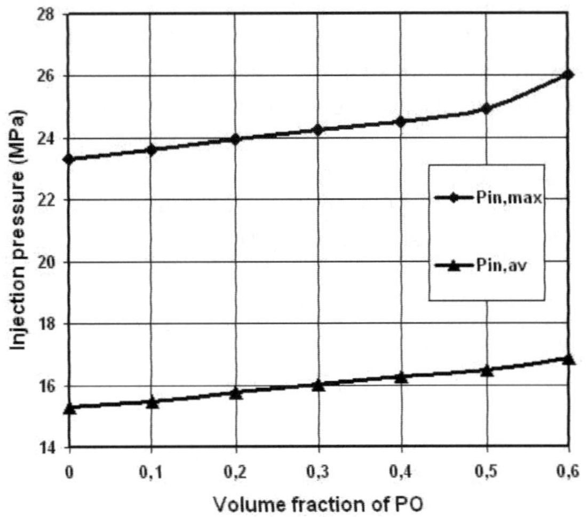

Figure 5. Dependence of the average $P_{in,av}$ and maximum $P_{in,max}$ injection pressures on volume fraction PO g_{po} at a constant amount of the heat generated

The total duration of injection under these conditions remained virtually unchanged. Analysis of Figures 4b, 3b and 5 shows, that the mass and volume injection of the mixture increases as the mass flow of the mixture increases less than when maintaining a constant active stroke of the plunger (compare Fig. 1b). The active plunger stroke is reduced less when maintaining a constant mass of the cyclic flow of the mixture (compare Fig. 3a). The average and maximum injection pressure increase less while maintaining a constant cyclic feed of the mixture (compare Fig. 2b).

In our opinion, for the practical use of the most valuable results presented on the calculations of fuel injection performed while maintaining a constant the amount of heat introduced by the fuel into the combustion chamber. The change in the mass and volumetric flow of mixed fuel and adjusting injection pump setting – the active stroke of the plunger when changing the volume fraction g_{po} of palm oil in the mixture, for this case are shown in table 3.

Table 3. The variation of the adjusting parameters of fuel injection pump when changing the volume fraction of palm oil

Variation of adjusting parameters	Units	D100	PO10	PO20	PO30	PO40	PO50	PO60
Mass cyclic flow of the fuel	g	0	0,0006	0,0013	0,0019	0,0025	0,0032	0,0038
mixture	%	0	1.06	2.12	3.18	4.24	5.3	6.36
Volumetric cyclic flow of the fuel	cm^3	0	0.0002	0.0003	0.0005	0.0008	0.001	0.0013
mixture	%	0	0.22	0.48	0.79	1.07	1.41	1.78
Active stroke of	mm	0	0.02	0.03	0.038	0.045	0.054	0.056
the plunger	%	0	1.04	1.56	1.97	2.33	2.8	2.9

4. Conclusion

Thus, the use of fuel blends, having palm oil as one of the components in operating diesel equipment with injection system type direct injection, does not require additional adjustment of fuel injection pump to maintain constant power of the operating diesel engine. As such the required variations in the change of *volumetric* cyclic flow and the change in active stroke of the plunger in accordance with changing the proportion of palm oil in the mixture up to 60% do not exceed 3 %. For diesels engines using a multi-fuel supply system, the deviations of up to 5% [22] at the nominal cyclic flow are admitted. However, these deviations for the injection system fuelling the operating diesel engine in this study are within acceptable limits, which allow the use of mixed fuel based on palm oil.

References

[1] Devyanin S N, Markov V A, Semenov V G 2008 Rastitelnyie masla i topliva na ih osnove dlya dizelnyih dvigateley (Vegetable oils and fuels on their basis for diesel engines). *FGOU VPO MGAU, Moskva*

[2] Lukanin V N et al. 2007 Dvigateli vnutrennego sgoraniya. Kniga 1. Teoriya rabochih protsessov (The internal combustion engines. Book 1. The theory of working processes). *Vyisshaya shkola, Moskva*

[3] Markov V A, Gayvoronskiy A I, Grehov A V et al 2008 Rabota dizeley na netraditsionnyih toplivah (Operation of diesel engines on alternative fuels). *Legion-Avtodata, Moskva*

[4] Aleksandrov A A; Arharov I A; Markov V A et al 2012 Alternativnyie topliva dlya dvigateley vnutrennego sgoraniya. (Alternative fuel for internal combustion engines). *OOO NITs «Inzhener», OOO «Oniko-M», Moskva*

[5] Valeho P R et al 2014 Sravnitelnyie ispyitaniya alternativnyih topliv dlya dizelnyih dvigateley

(Comparative tests of alternative fuels for diesel engines). *Vestnik MGTU im. N. E. Baumana. Ser. "Mashinostroenie"* 6:596.

[6] Vedruchenko V R, Kraynov V V, Litvinov P V 2016 Vliyanie svoystv raznosortnyih topliv dlya dizeley na harakteristiki toplivopodachi (The influence of the properties of different sorts of fuels for diesel engines on the characteristics of the fuel supply). *Vestnik SibADI, Omsk.* 2(48):44-49.

[7] Priyandaka A, Valeho P I et al 2003 Eksperimentalnoe opredelenie kineticheskih konstant vosplameneniya rastitelnyih topliv v usloviyah DVS (Experimental determination of kinetic constants of ignition of vegetative fuels in the conditions of the internal combustion engine). *Vestnik Rossiyskogo universiteta druzhbyi narodov: Inzhenernyie issledovaniya* 1:29-31

[8] Markov V A 2014 Vliyanie sostava smesevogo biotopliva na parametryi protsessa vpryiskivaniya topliva v dizele (The influence of composition of mixed biofuels on process parameters of injection of fuel in a diesel engine). *Traktoryi i selhozmashinyi* 12:3-9

[9] Grehov L V, Gabitov I I, Negovora A V 2013 Konstruktsiya, raschet i tehnicheskiy servis toplivopodayuschih sistem dizeley (Design, calculation and technical services fuel supply systems of diesel engines). *Legion-Avtodata, Moskva*

[10] Moshin R, Majid Z A , Shiman A H, Nasri N S, Sharer Z 2014 Effect of biodiesel blends on engine performance and exhaust emission for diesel dual fuel engine. *Energy conversion and management* 88:821-828

[11] Januan J, Ellis N 2010 Perspectives on biodiesel as sustainable fuel. *Renew sustain Energy Rev* 14:1312-1320

[12] Ziejewski M, Kaufman K 1983 Vegetable oils as a potential alternate fuel in direct injection diesel engines. *SAE technical paper.*

[13] Martinez G, Sanchez N, Enginar J M, Gonzalez J F 2014 Fuel properties of biodiesel from vegetable oils and oil mixtures. Influence of methyl esters distribution. *Boimass bioenergy* 63:22-32

[14] Mosarof M H et al. 2015 Implementation of palm biodiesel based on economic aspects, performance ,emission and wear characteristics. *Energy conversion and management* 105:617-629

[15] Ndayishimiye P, Mohamed Tazerout 2011 Use of palm oil-based biofuel in the international combustion engines: Performances and emissions characteristics. *Energy* 361:790-796

[16] Sanjid A et al. 2013 Impact of palm, mustard, waste cooking oil and calophyllum inophyllum biofuels on performance and emission of CI engine. *Renewable and sustainable energy reviews* 27:664-682

[17] Yusaf T F, Yousif B F, Elawad M M 2011 Crude palm oil fuel for diesel-engines: experimental and ANN simulation approaches. *Energy* 36:4871-4878

[18] Rashid M M et al. 2016 Performance and emission characteristics of a diesel engine fueled with palm, jatropha and moringa oil methyl ester. *Industrial crops and products* 79:70-75

[19] Azad A, Khan M, Ahasan T, Ahmed S 2015 Energy scenario: production consumption and prospect of renewable energy in Australia. *Power Energy* Eng 2:19

[20] Azad A, Uddin S A 2013 Performance study of a diesel engine by first generation biofuel blends with fossil fuel: an experimental study. *Renew sustain Energy* 5:013-118

[21] Shy E G (1993) Diesel fuel from vegetable oils: status and opportunities. *Biomass bioenergy* 4:227-242.

[22] GOST 10578-95 Nasosyi toplivnyie dizeley. Obschie tehnicheskie usloviya (Diesel engine fuel pumps. General specifications). *Izdatelstvo standartov, Moskva.*

IOP Conf. Series: Materials Science and Engineering **386** (2018) 012015 doi:10.1088/1757-899X/386/1/012015

Improvement of bogie tracks for forestry wheeled vehicles

V E Klubnichkin, E E Klubnichkin, V N Naumov

Bauman Moscow State Technical University, 5/1, 2-ya Baumanskaya st., Moscow, 105005, Russia

E-mail: vklubnichkin@gmail.com

Abstract. The article is concerned with development of a bogie track for forestry wheeled vehicles featuring extended service life, better grip and lower ground pressure as compared with the counterparts. The conditions and results of the experimental studies of wheeled forestry vehicles with and without different anti-skid tracks are described in the article.

Introduction

Today contemporary logging methods are frequently used by the logging companies both in Russia and all over the world. The world practice demonstrates that the cut-to-length method employs machines, including a harvester and a forwarder [1, 2, 3].

Figure 1 shows an example of forestry wheeled vehicles equipped with bogie tracks.

Figure 1. Forestry vehicles which may be equipped with bogie tracks

These vehicles show good efficiency thanks to modern engineering equipment and automated control system. However, the vehicles suffer from one fatal flow. They are wheeled so it makes them inefficient on swampy, running soils and under subzero temperatures in such conditions as snow and glass [4, 5, 6]. Practise has shown that these vehicles are fitted with standard bogie tracks manufactured by Olofsfors, Clark and other companies.

The main feature of the bogie tracks is that the tracks distribute load over greater area and do not concentrate it on two small flat spots. Brush wood, roots, etc. often lay on the ground and help supporting a tracked machine. Tires usually press the ground and pack it.

The advantage of using properly selected bogie tracks is soil damage protection; less ground pressure and better ground clutch; less soil compacting: better propulsive thrust; reduced fuel consumption; increased cargo capacity; better stability during loading and unloading; better movement stability.

Soil is one of the main components of the forest eco-system serious damage of which can have a negative impact on vegetatio. Soil formation process takes time and soil structure may be weak.
The standard bogie tracks represent tracks inter-connected with a ring, i.e. open joint (Fig. 2).

Figure 2. Standard bogie tracks

An open joint is out of order very quickly while contacting with soil abrasive elements such as sand, ground, etc.

After a while a track gets extended and becomes sagged and deformed under pressure of a vehicle load and ground deformation. It occurs when the track is installed on the balance truck, the second wheel always suffers from increased travel resistance, i.e uphill movement (Fig. 3) [7, 8, 9].

Figure 3. Standard bogie track sagging

Taking into consideration a large number of vehicles with different load and use methods, it is very difficult to specify exact track service life. Great tension and load affecting the tracks should also be considered. The average service life of a standard track with an open joint is around 3 thousand moto hours.

The vehicle which isn't equipped with the tracks has negative effect on the soil resulting in packing down to critical values. The ground pressure is very high and is up to standard of around 75 kPa.

Considering the aforesaid, we have decided to improve joint elements of standard bogie tracks in order to Decrease ground pressure; Reduce travel resistance factor; Reduce fuel consumption; Improve track service life; Improve maintainability.

1. The developed enclosed metal-rubber mountings in the track.

We have reviewed the existing technical solutions in this field and come to a conclusion that the most rational option to solve the task would be to use enclosed metal-rubber mountings in the track (Fig. 4).

IOP Conf. Series: Materials Science and Engineering **386** (2018) 012015 doi:10.1088/1757-899X/386/1/012015

Figure 4. Metal-rubber mounting

This design is based on a silent block (Fig. 5). The rubber bushings and guides are cured and pressed fit into the joint housing on the press, and then the ready joints are assembled into the links on the track.

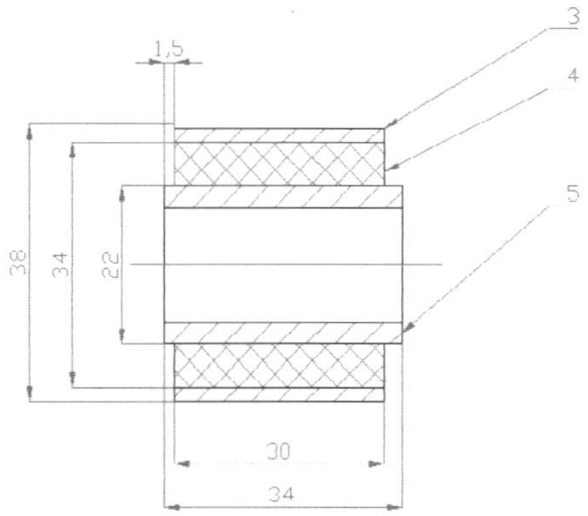

Figure 5. Silent block

3. Metal bushing; 4. Elastic insert (rubber, grade IRP-1357); 5. Metal bushing

Figure 6 shows a bogie track being developed

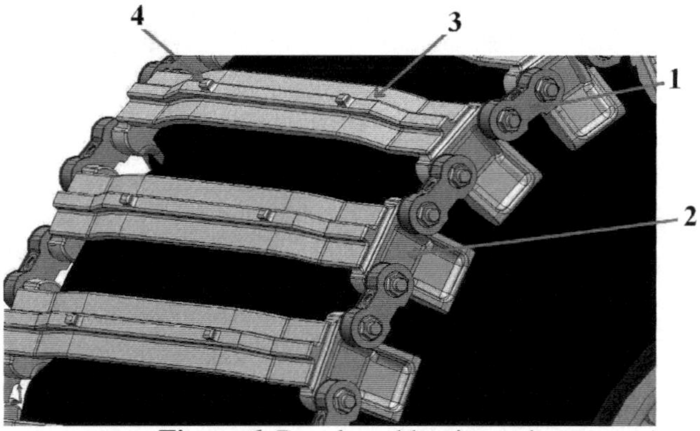

Figure 6. Developed bogie track

1. Metal-rubber mounting; 2. Side support; 3. Track; 4. Grouser

The strengthened use and use with varying wheel load gives extra tension to the tire bearing capacity and the pressure value, which is especially important for the forestry vehicles [10]. To improve tire strength and service life it is necessary to follow the manufacturers' recommendations since the track inner part remained unchanged, so the tire pressure recommendations (Table 1) remain the same.

IOP Conf. Series: Materials Science and Engineering **386** (2018) 012015 doi:10.1088/1757-899X/386/1/012015

Table 1. Tire pressure recommendations

Tires			
Tire size	Tire brand	Model of tires	Maximum pressure
600/50x 22.5	Nokian	ELS L-2(1 6)	4.3 bar/62 psi
	Nokian	TRS LS-2(16)	4.3 bar/62 psi
	Trelleborg	T404	4.0 bar/58 psi
	Trelleborg	T423	4.0 bar/58 psi
	Trelleborg	T428 143/150	4,0 bar/58 psi
	Trelleborg	T428 149/156	5.0 bar/73 psi
650/45 x 22.5	Trelleborg	T422 150	5,0 bar/73 psi
700/45 x 22.5	Nokian	ELS L-2(1 6)	3.9 bar/57 psi
	Nokian	TRS LS-2(16)	3.9 bar/57 psi
700/40 x 22.5	Trelleborg	M404	4.0 bar/58 psi
	Trelleborg	T423	4.0 bar/58 psi
710/40 x 22.5	Nokian	FK F	3.9 bar/57 psi
	Trelleborg	T428 145/152	4,0 bar/58 psi
	Trelleborg	T428 151/158	5.0 bar/73 psi
600/55 x 26.5	Nokian	ELS L-2(1 6)	4.6 bar/67 psi
	Nokian	ELS L-2(20)	5.5 bar/80 psi
	Nokian	TRS LS-2(16)	4.6 bar/67 psi
	Nokian	FK F (16)	4.6 bar/67 psi
	Nokian	FK F (20)	5.5 bar/80 psi
	Trelleborg	T421	5.0 bar/73 psi
	Trelleborg	T428 160/167	6.0 bar/87 psi
	Trelleborg	T422 154/161	5.0 bar/73 psi
	Firestone	Forestry EL	3.9 bar/56 psi
620/55 x 30.5	Trelleborg	T428 158/165	5.0 bar/73 psi
650/65 x 26.5	Nokian	ELS L-2	5.5 bar/80 psi
650/60 x 26.5	Trelleborg	T428 161/168	5.0 bar/73 psi
600/60 x 30.5	Trelleborg	T421	5.0 bar/73 psi
700/50 x 26.5	Nokian	ELS L-2(1 6)	4.6 bar/67 psi
	Nokian	ELS L-2(20)	5.5 bar/80 psi
	Nokian	TRS LS-2(16)	4.6 bar/67 psi
	Nokian	TRS LS-2(20)	5.5 bar/80 psi
	Trelleborg	T423	5.0 bar/73 psi
	Firestone	Forestry EL	4.6 bar/66 psi
710/45 x 26.5	Nokian	FK F (16)	4.6 bar/67 psi
	Nokian	FK F (20)	5.5 bar/80 psi
	Trelleborg	T428 151/158	4,5 bar/65 psi
	Trelleborg	T428 163/170	6.0 bar/87 psi
750/50 x 26.5	Trelleborg	T428 163/170	5,0 bar/73 psi
750/50 x 26.5	Trelleborg	T428 170/177	6.0 bar/87 psi
700/50 x 30.5	Trelleborg	T423	5.0 bar/73 psi
750/55 x 26.5	Nokian	ELS L-2(20)	5.5 bar/80 psi
	Nokian	TRS L-2(20)	5.5 bar/80 psi
	Nokian	FK F(20)	5.5 bar/80 psi
750/45 x 30.5	Trelleborg	T428 169/176	6.0 bar/87 psi
800/40 x 26.5	Nokian	FK F(20)	5.0 bar/73 psi
	Trelleborg	T423	5.0 bar/73 psi

IOP Conf. Series: Materials Science and Engineering **386** (2018) 012015 doi:10.1088/1757-899X/386/1/012015

We have made torsion and cranking resistance designs for the silent block to reduce ground pressure. This allowed strength of the metal-rubber mounting would strive to turn the track towards the ground contacting area [11]. This also solves the task of reducing travel resistance as the ground contacting area enlarges; the metal-rubber mounting strength constantly levels the ground contacting area and the second wheel of the balance truck moves with least travel resistance that should significantly reduce fuel usage. The bogie tracks with metal-rubber mountings were tested on timber harvesting machine Ponsse Beaver in Murashinsky logging company in Kirov region.

2. Test conditions

The tests were performed in winter under the conditions when the tire clutch on the snow surface is minimum and it must be increased to improve running performance of the wheeled vehicles. All winter tests were performed under air temperature from -2 to +5°C in three backgrounds: virgin snow 16 – 25 cm deep and density of 0.2 – 0.26 g/cm^3, snow road with depth of snow layer pressing depth of 9 – 15 cm with density of 0.45 – 0.51 g/cm^3 and snow-iced compacted road with upper layer density of 0.64 – 0.66 g/cm^3. During testing the snow surface temperature was from -2.5 to +0°C. The selected routes used for runs were horizontal without visible inclines and rises. Due to dry winter we could not find sections with deeper snow for tests. During the tests the air and snow were measured with thermometer. The snow density was detected by taking samples by a gage glass on measuring sections of the route. This snow was melted and then a measuring tube was used to determine the volume of the water and hence the weight of the scow sample.

In spring the tests were performed on dry (W = 15-17%) and wet (W = 28-33%) loamy roads.

The moisture and density of soil were determined with the Kovalev meter.

During the test pulling characteristics of the vehicle with two types of bogie tracks (standard and with metal-rubber mountings) as well as without tracks were measured. The characteristics were measured during steady movement on the sections 50 meter long. Ehe tested vehicle carried additional load of 2000 kg.

The pulling loads were created by tractor K-703. The loads were measured with pulling stain-gage link inserted into the towing cable. Wheel spinning of the tested vehicle was determined by the results of measurement of wheels' turns and track-measuring wheel turns with a sensing probe. The measured values were recorded with a decoder. The recording equipment was located in the cab of tractor K-703.

The tested vehicle had tracks on the tandem dolly only.

3. Test results

Table 2 gives values of clutch coefficients of the vehicle with standard tracks, developed tracks and without tracks.

Table 2.

Background	Grip coefficient		
	Developed tracks	Olofsfors ECO-TRACK	Without tracks
Virgin snow	0.59	0.39	0.34
Snow road	0.53	0.49	0.30
Snow-iced road	0.57	0.50	0.29
Wet loamy road	0.62	0.55	0.53
Dry loamy road	0.73	0.68	0.72

The given figures show that the vehicle equipped with developed tracks features better pull during spinning by 40 - 50 %. The developed bogie tracks are more efficient than the standard ones by 14 - 18 %.

IOP Conf. Series: Materials Science and Engineering **386** (2018) 012015 doi:10.1088/1757-899X/386/1/012015

4. Conclusion

Using bogie tracks in the structure as joint elements of the metal-rubber mountings makes it possible to reduce ground pressure, travel resistance factor, fuel consumption, improve track service life and their maintainability.

Working on a very soft soil it is necessary to use metal-rubber mounting along with tracks of a special form which protects soil and its subsurface against damage.

The given data shows that the vehicle equipped with developed tracks features better pull during spinning by 40 - 50 %. The developed tracks are more efficient than the standard ones by 14 - 18 %.

References

[1] Voskoboynikov I, Kondratyuk V, Krylov V, Kondratyuk D, Klubnichkin E 2013 Multioperational forest machines. *Monograph. Vol. 1. Pages 480.*

[2] Ksenevich I, Goberman V, Goberman L 2003 Ground trailer transport systems. *Vol. 3. Encyclopedia. – Moscow, Engineering. Pages 788.*

[3] Kotikov V, Eremeev N, Erkhov A 2004 Forestry and skidding machines. *Moscow, Forest industry,. Pages 336*

[4] Eliasson L, Wasterlund I, 2007 Effects of slash reinforcement of strip roads on rutting and soil compaction on a moist fine-grained soil. *Forest Ecology and Management 252 (1-3), pp. 118–123.*

[5] Klubnichkin V 2010 Evaluation of the influence of external conditions on forest machines *Moscow State Forest University Bulletin – Lesnoy vestnik. №6, pp. 119-123.*

[6] Ageykin Ya 1972 Rough terrain wheels and dual propellers. *Moscow, Engineering. Pages 184.*

[7] Ansorge D, Godwin R, 2007 The effect of tyres and a rubber track at high axle loads on soil compaction, part1: Single axle-studies. *Biosystems Engineering 98 (1), pp. 115–126.*

[8] Eliasson L 2005 Effects of forwarder tyre pressure on rut formation and soil compaction. *Journal of Silva Fennica 39 (4), pp. 549–557.*

[9] Klubnichkin V, Kotikov V, Klubnichkin E 2010 General technique of research passableness wheeled vehicles with a bogie axle, equipped with tracks chains. *Natural and engineering sciences. № 3, pp. 327-334.*

[10] Egorov L, Klubnichkin V, Klubnichkin E 2013 Wheels and tires forest machines. *Moscow, MSFU, Pages 36.*

[11] Kotikov V, Klubnichkin E 2010 On the problem of the method research of loading tracked chains of wheeled vehicles with Bogies. *Natural and engineering sciences. № 3, pp. 321-326.*

IOP Conf. Series: Materials Science and Engineering **386** (2018) 012016 doi:10.1088/1757-899X/386/1/012016

Interaction between elements of the track ground contacting area with the soil at curvilinear motion of the timber harvesting machine

V E Klubnichkin[1], E E Klubnichkin[1], G O Kotiev[1], S A Beketov[1], V S Makarov[2]

[1] Bauman Moscow State Technical University, 5/1, 2-ya Baumanskaya st., Moscow, 105005, Russia
[2] Nizhny Novgorod State Technical University n.a. R.E. Alekseev, Minin st., 24, 603950, Nizhny Novgorod, Russia

E-mail: vklubnichkin@gmail.com

Abstract. This article describes interaction between the tracks of the timber harvesting machine and the soil. It concerns the research of interaction of the elements of the ground contacting area of the tracks with the soil at curvilinear motion of the tracked timber harvesting machine. The expressions was found to determine relationship between the value of the longitudinal and lateral travel of an element of the track ground contacting area and the traveltime parameters of motion of the tracked timber harvesting machine and in turn characterize kinematics of the elements of the track ground contacting area at curvilinear motion of the machine.

1. Introduction

It is impossible to accomplish the task set by the development strategy of the timber complex of the Russian Federation until 2020 to increase timber harvesting up to 294 mln. cu.m. without providing the timber harvesting enterprises with a complex of machines and equipment ensuring efficient timber harvesting, including in the severe weather conditions which prevail in most forest areas (weak soil, heavily waterlogged area, deep snow, etc.).

It is advisable to use the tracked machines on the waterlogged cutting areas, steep slopes, in deep snow, multi-storeyed wood, cross-country full of obstacles in the form of stubs, stones and fallen trees, where the performance and traction power of the wheeled vehicles are insufficient. The advantage of the tracked running gear is even more evident on the ice surface [1, 2, 3, 4].

In this article a model of interaction between elements of the track ground contacting area with the soil is described. The kinematics of the elements of the ground contacting areas of the tracked timber harvesting machine at its curvilinear motion is examined

2. Model of interaction between elements of the track ground contacting area with the soil

The analytical description of the track-soil interaction involves considerable difficulties, though the physical and mechanical properties of soils have been well studied today. However, application of the soil mechanics laws to describe the track-soil interaction is not acceptable due to dynamic loading of the track ground contacting area elements on the soil.

IOP Conf. Series: Materials Science and Engineering **386** (2018) 012016 doi:10.1088/1757-899X/386/1/012016

Due to the above, experimental data which allows to find out some common patterns are used to describe the process of the track-soil interaction [5, 6, 7]. Let us examine interaction between the elements of the track ground contacting area with the soil using those patterns.

To evaluate this interaction let us take a track-related triangular coordinate system which zero coincides with the track centre-of-mass (Fig. 1).

During the curvilinear motion of the tracked timber harvesting machine, each element of the track ground contacting area carries out a complex motion, gets displaced on the soil in longitudinal and lateral direction relative to the machine rolling axis [8, 9]. While examining the track-soil interaction, we assume that the track travel on the soil at a specific time coincides with the direction of the absolute velocity of the track centre -of-mass travel. We can approximately accept that the track centre -of-mass coincides with the geometric centre of its surface.

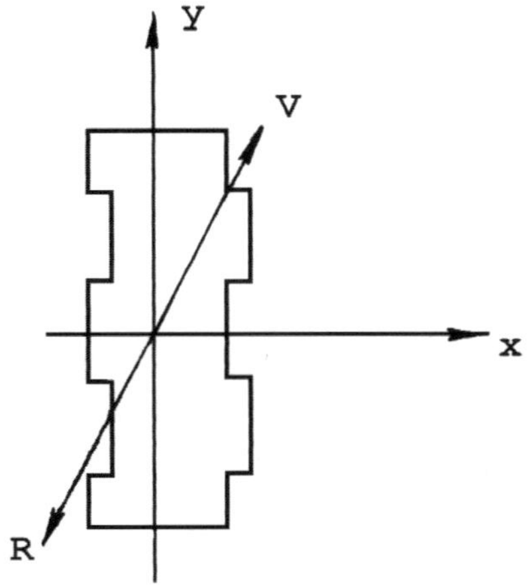

Figure 1. Coordinate system to analyze the track-soil interaction

Let us suppose the track to be loaded with force *Q* (normal track-on-soil load) which is normal to its surface. When the track travels relative to the soil, this causes a response force that we mark as *R*. This force is proportional to the normal load:

$$R = \mu Q. \tag{1}$$

The proportionality factor μ is called a factor of the track-soil interaction.

Let us assume that the force of the track-soil interaction is opposite in direction to the instantaneous sliding velocity of the centre -of-mass of the track relative to the soil *V* (Fig. 1).

The track-soil interaction force *R* in general depends on the value and direction of travel of the track relative to the soil because the track ground contacting area has a complex shape and the forces resisting its travel in longitudinal and lateral direction are different.

Therefore, with the same value of track travel relative to the soil σ and constant value of the normal track-on-soil load *Q*, the reaction force *R* will be different depending on the track sliding direction (there will be different value of coefficient μ).

Let us call the hodograph of force *R* at *Q*=1 the hodograph of the track-soil interaction coefficient μ.

The results of experiments showed that this hodograph practically has the shape of ellipsis and may be described in the following equation:

$$\mu = \frac{\mu_x \cdot \mu_y}{\sqrt{\mu_x^2 \cdot \sin^2 \alpha + \mu_y^2 \cdot \cos^2 \alpha}}, \tag{2}$$

where μ_x, μ_y is a value of the coefficient of the track-soil interaction at (separate) travel of the track by value σ in longitudinal and lateral directions; α is an angle between direction of the track displacement and longitudinal axis of the tracked timber harvesting machine.

Unequal soil resistance to the track displacement in longitudinal and lateral directions is taken into consideration with use of the anisotropy coefficient λ:

$$\lambda = \mu_{y\,\max} / \mu_{x\,\max}. \quad (3)$$

3. Kinematics of the elements of the track ground contacting areas at curvilinear motion of the tracked timber harvesting machine

Let us examine the kinematics of the elements of the ground contacting areas of the tracked timber harvesting machine at its curvilinear motion (Fig. 2).

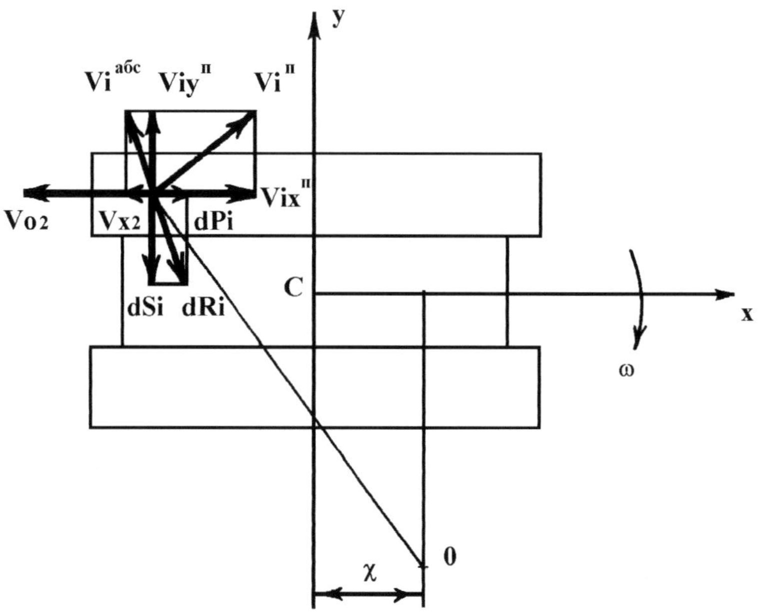

Figure 2. Kinematics of the elements of the track ground contacting areas at curvilinear motion of the tracked timber harvesting machine

Assume that point O is an instantaneous center-of-turn. Select *i*-nd element on the ground contacting area of the leading track. It will participate in two motions: translational (around the instantaneous center-of-turn) at a rate V_i^n and relative (track return relative to the body) at a rate V_{02}. (Further all parameters related to the leading track will be marked with subindex "2", and to the retreating track - with subindex "1".) Let us break down translational velocity V_i^n into longitudinal V_{ix}^n and lateral V_{iy}^n components. The difference between relative velocity V_{02} and longitudinal component V_{ix}^n will give slipping velocity V_{X2} of the leading track relative to the soil. The vector sum of velocities V_{iy}^n and V_{X2} will give an absolute velocity of element displacement $V_i^{абс}$ which the direction of the soil absolute response dR_i depends on. Let us assume the following: vector of soil absolute response coincides with the vector of absolute displacement of the *i*-nd element of the track ground contacting area relative to the soil but is opposite in direction. Considering this assumption, the soil absolute response dR_i may be broken down into two components: lateral - relative to the axis of the tracked timber harvesting machine (elementary turn resistance force dS_i) and longitudinal one which is directed along this axis (elementary traction power dP_i).

In view of the above, the soil response acting on each element of the track ground contacting area may be expressed with a reasonable degree of accuracy through longitudinal and lateral displacement of the element of the ground contacting area relative to the soil.

To determine the values of longitudinal and lateral travel of each element of the track ground contacting area over the soil, let us examine their kinematics at the curvilinear motion of the tracked timber harvesting machine.

At each point of time of the plane curvilinear motion of the tracked timber harvesting machine, its travel as a solid object can be replaced with rotating motion relative to the instantaneous center-of-turn *O* (Fig. 3).

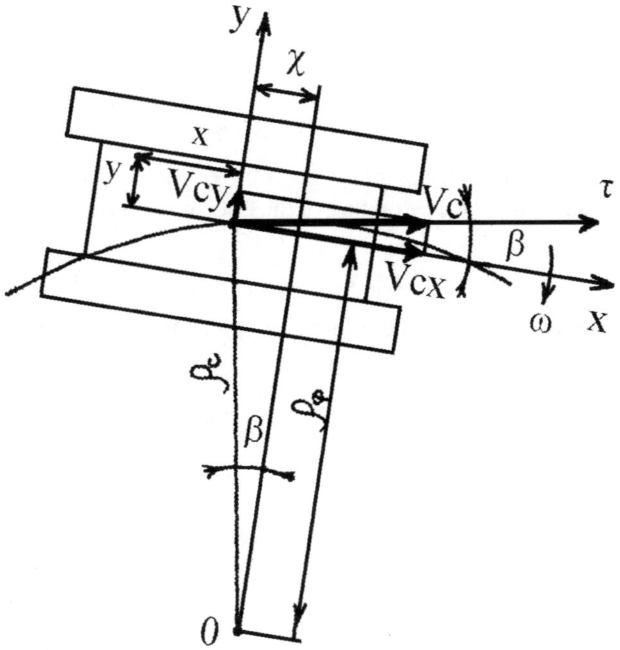

Figure 3. Kinematics of the curvilinear motion of the tracked timber harvesting machine relative to the instantaneous center-of-turn

Direction of axis *x* coincides with the direction of the machine longitudinal axis.

Take ω as an angular rate of rotation of the tracked timber harvesting machine relative to the instantaneous center-of-turn *O*, record the expression of motion velocity of its center-of-mass:

$$V_C = \omega \cdot \rho_C \cdot \tag{4}$$

Break down the center-of-mass velocity of the tracked timber harvesting machine V_C into longitudinal V_{CX} and lateral V_{CY} components:

$$V_{CX} = V_C \cdot \cos\beta = \omega \cdot \rho_C \cdot \cos\beta = \omega \cdot \rho_\Phi = \omega \cdot \sqrt{\rho_C^2 - \chi^2}; \tag{5}$$

$$V_{CY} = V_C \cdot \sin\beta = \omega \cdot \rho_C \cdot \sin\beta = \omega \cdot \chi, \tag{6}$$

where ρ_C is a radius of curvature of trajectory of the machine center-of-mass;

χ is a projection of instantaneous center-of-turn to axis *x*;

ρ_Φ is an actual turn radius;

β is an angle between the tangent to the required motion trajectory and longitudinal axis of the tracked timber harvesting machine.

It is evident that

$$\beta = \arcsin(\chi / \rho_C). \tag{7}$$

The projection of instantaneous velocity of the body arbitrary point A with coordinates x and y similarly with velocities V_{CX} and V_{CY} shall be determined with the following expressions:

$$V_{AX} = \omega \cdot (\sqrt{\rho_C^2 - \chi^2} + y); \tag{8}$$
$$V_{AY} = \omega \cdot (\chi - x).$$

The tracks are not rigidly connected with the body of the tracked timber harvesting machine in longitudinal direction in relative motion and have rigid connection in the lateral direction. For this the lateral velocity of travel of the elements of the ground contacting areas shall be determined with the following expression:

$$V_{y2}(x) = V_{y1}(x) = \omega \cdot (\chi - x), \tag{9}$$

where $V_{y2}(x)$ and $V_{y1}(x)$ are the lateral velocities of travel of the point with coordinate x belonging to the ground contacting area of the leading and retreating tracks, respectively.

The elementary move of the point belonging to the track ground contacting area shall be defined by the following dependence:

$$d\sigma_{y2} = d\sigma_{y1} = \omega \cdot (\chi - x) \cdot dt_k, \tag{10}$$

where k is the time of contact of this point of the ground contacting area with the soil.

For the leading track:

$$t_{k2} = [(l_1 + t_\Gamma / 2) - x] / V_{02}. \tag{11}$$

Similar for the retreating track:

$$t_{k1} = [(l_1 + t_\Gamma / 2) - x] / V_{01}, \tag{12}$$

where V_{02}, V_{01} are respective velocities of the leading and retreating tracks' turn relative to the tracked timber harvesting machine body.

It is evident that

$$(l_n - t_\Gamma / 2) \le x \le (l_1 + t_\Gamma / 2), \tag{13}$$

where l_n is a distance between the center-of-mass of the tracked timber harvesting machine to the last support roller.

The fill displacement of the element of the track ground contacting area over the soil in lateral (relative to the longitudinal axis of the tracked timber harvesting machine) direction shall be determined by the following expressions:

$$\sigma_{y2}(x) = \int_0^{t_{k2}} \omega(t) \cdot (\chi - x) \cdot dt_{k2}; \tag{14}$$
$$\sigma_{y1}(x) = \int_0^{t_{k1}} \omega(t) \cdot (\chi - x) \cdot dt_{k1};$$

at ω = const

$$\sigma_{y2}(x) = \omega \int_0^{t_{k2}} (\chi - x) \cdot dt_{k2}; \tag{15}$$

$$\sigma_{y1}(x) = \omega \int_0^{t_{k1}} (\chi - x) \cdot dt_{k1}.$$

In formulas (15) let us go to variable x instead of t_k. In accordance with expressions (11) and (12) we arrive at:

$$dt_{k1}=-dx/V_{01};$$
$$dt_{k2}=-dx/V_{02}.$$

Consequently, expressions (15) may be recorded in the following form:

$$\sigma_{y2}(x) = (\omega / V_{02}) \cdot \int_x^{l_1+t_\partial/2} (\chi - x)dx; \tag{16}$$

$$\sigma_{y1}(x) = (\omega / V_{01}) \cdot \int_x^{l_1+t_\partial/2} (\chi - x)dx.$$

Having calculated integrals in expressions (16), we arrive at:

$$\sigma_{y2}(x) = [\omega/(2 \cdot V_{02})] \cdot \left((\chi - x)^2 - (\chi - l_1 - t_\partial/2)^2\right); \tag{17}$$

$$\sigma_{y1}(x) = [\omega/(2 \cdot V_{01})] \cdot \left((\chi - x)^2 - (\chi - l_1 - t_\partial/2)^2\right).$$

Let us determine longitudinal travel of the elements of the ground contacting areas over the soil:

$$\sigma_{x2} = V_{x2} \cdot t_{k2}; \tag{18}$$

$$\sigma_{x1} = V_{x1} \cdot t_{k1}, \tag{19}$$

where V_{x2} is a slipping velocity of the leading track; V_{x1} is a slipping velocity of the retreating track. But for V_{x2} and V_{x1} we can record:

$$V_{x2} = V_{x2}^n - V_{02}; \tag{20}$$

$$V_{x1} = V_{x1}^n - V_{01}, \tag{21}$$

where V_{x2}^n and V_{x1}^n are translational velocities of the center-of-mass of the track line that may be determined as follows:

$$V_{x2}^n = \omega \cdot \left(\sqrt{\rho_C^2 - \chi^2} + B/2\right) = V_{CX} + \frac{\omega \cdot B}{2};$$

$$V_{x1}^n = \omega \cdot \left(\sqrt{\rho_C^2 - \chi^2} - B/2\right) = V_{CX} - \frac{\omega \cdot B}{2}.$$

Having substitute expressions (11), (12), (20), (21) в (18) and (19), we arrive at:

$$\sigma_{x2}(x) = \left(\omega \cdot (\sqrt{\rho_C^2 - \chi^2} + B/2) - V_{02}\right) \cdot (l_1 + t_\Gamma/2 - x)/V_{02}; \tag{22}$$

$$\sigma_{x1}(x) = \left(\omega \cdot (\sqrt{\rho_C^2 - \chi^2} - B/2) - V_{01}\right) \cdot (l_1 + t_\Gamma/2 - x)/V_{01}. \tag{23}$$

4. Conclusion

Thus the expressions determined relationship between the value of the longitudinal and lateral travels of an element of the track ground contacting area and the travel time parameters of motion of the tracked timber harvesting machine and in its turn characterize kinematics of the elements of the track ground contacting area at curvilinear motion of the machine was found

IOP Conf. Series: Materials Science and Engineering **386** (2018) 012016 doi:10.1088/1757-899X/386/1/012016

References

[1] Bekker M G 1969 Introduction to Terrain-Vehicle Systems. *The University of Michigan Press, Ann Arbor.*
[2] Wong J Y 2001 Theory of Ground Vehicles. *third ed. John Wiley & Sons, NY, ISBN 0-471- 35461-9.*
[3] Wong J Y 2009 Terramechanics and Off-Road Vehicle Engineering. *2nd Edition. Pages 488.*
[4] Kotikov V M 1995 Impact harvesting machines on forest soils. *Dissertation of the doctor of technical sciences. Moscow state forest university, Pages 214.*
[5] Beketov S 2017 The theory of the controlled motion of tracked vehicles. *Publishing house Bauman MSTU. Pages 125.*
[6] Klubnichkin V, Klubnichkin E, Kotiev G 2017 Developing a method of theoretical evaluation of the tracked timber harvesting machine undercarriage element loading. *Proceedings of the 19th International & 14th European-African Regional Conference of the ISTVS*
[7] Klubnichkin E, Klubnichkin V 2017 Theoretical foundations of preliminary assessment of the impact of the timber harvesting vehicles on forest soils in the course of logging. *Proceedings of the 19th International & 14th European-African Regional Conference of the ISTVS*
[8] Said Al-Milli, Lakmal D, Kaspar A 2010 Track-terrain modelling and traversability prediction for tracked vehicles on soft terrain. *Journal of Terramechanics, Vol. 47, Issue 3, pp. 151-160.*
[9] Bodin A 1999 Development of a tracked vehicle to study the influence of vehicle parameters on tractive performance in soft terrain. *Journal of Terramechanics. Vol. 36, Issue 3, 167-181.*

Light commercial vehicles designing with considering end-of-life vehicles recycling requirements

S A Gagunov[1], A M Groshev[2], D.M. Porubov[2], V.V. Belyakov[2], D.V. Zeziulin[2]

[1]Co LLC "United Engineering Centre", Lenin ave., 88, Nizhny Novgorod, 603004, Russian Federation
[2]Nizhny Novgorod State Technical University n.a. R.E. Alekseev, Minin str., 24, Nizhniy Novgorod, 603950, Russian Federation

E-mail: pavel.beresnev@nntu.ru

Abstract. The issues of utilization of vehicles at the end of their life cycle are the most important from the ecological point of view. Foreign experience in the field of utilization of vehicles also makes it possible to talk about the economic efficiency of this industry, but in order to achieve high results it is necessary to take into account the requirements for recycling at the early stages of product design. This article analyzes the domestic recycling industry. The main barriers in the field of utilization of commercial vehicles have been identified. A methodology for designing light commercial vehicles (LCA) is proposed, which satisfies the utilization rates. A design algorithm is proposed to take into account the requirements of engineering analysis and economic feasibility, which ensures product orientation for recycling. Calculation of recycling indicators based on a complete set of information on vehicle components is given. Dismantling of vehicle components as a key process in the utilization structure is considered. A comparative evaluation of the efficiency of dismantling of plastic products of GAZ Group commercial vehicles was carried out.

1. Introduction

A growing fleet of cars and, as a result, the growing burden on the environment poses the task of minimizing the negative impact of vehicles on the environment. One of the key areas of increasing environmental friendliness is the task of recycling end-of-life vehicles. At the same time, the degree of reuse of materials and components used in the manufacture of cars is essential for improving the energy efficiency and resource saving of the economy.

Almost all main participants of the motor transport complex are interested in the recycling of cars. Automakers receive significant benefits through the use of cheaper secondary raw materials, trucking companies and car owners from the state provide cash subsidies. The Ministry of Industry and Trade, through recycling, promotes renewal of the fleet of vehicles. Renewal of the fleet is also beneficial for the Ministry of Transport: due to recycling of vehicles, the transport system safety is increasing and the environment is improving [1].

Foreign countries have tremendous experience in the field of vehicles utilization, as a result this sphere of activity brings quite impressive income, which allows the industry to develop itself without significant government intervention. So in the countries of the European Union, according to the

IOP Conf. Series: Materials Science and Engineering **386** (2018) 012017 doi:10.1088/1757-899X/386/1/012017

European Association of Automobile Manufacturers, the profit from processing one car in the EU is about 340 euros [2].

2. Domestic experience in the field of utilization of vehicles

Unlike foreign countries, the domestic recycling industry is now in the initial stage of development. Along with the problems in the legal sphere and the lack of a unified system of utilization centers, there are difficulties in the utilization of vehicles, the design of which does not allow a qualitative and rapid implementation of the process of dismantling and subsequent processing of the product.

So the issue of utilization of light commercial vehicles of GAZ Group has a great importance in view of the fact that the company's products occupy about 70% of the total number of vehicles sold in this category on the domestic market. The adaptation of the design to recycling allows not only to meet environmental requirements in the domestic market, but also to contribute to the increase in the export potential of the product. For the most qualitative solution of the problem of motor transport utilization it is necessary, at the stage of designing new products, to put the possibility of maximal reuse and recycling of components and materials after the end of the operation phase of the car.

The solution of tasks related to the assessment of environmental friendliness and the development of methodologies for the design of vehicles, taking into account their subsequent disposal, involves various scientists and scientific organizations.

In the papers [3, 4] technological schemes for utilization of vehicle components and assemblies were considered, an analysis of foreign experience in the field of utilization was made, and the domestic regulatory framework was evaluated.

In the paper [5] the technique of an ecological estimation of cars in a full life cycle, considering stages of their manufacture, operation and recycling is presented.

In [6] the methodology of constructing advanced design of vehicles with their subsequent utilization is presented, which implies the use of a set of recommendations about the modular principles for the design of vehicle units and systems, the use of connections in the vehicle design that facilitate simplified disassembly of parts, and also sorting of materials, selection of structural materials in units and systems from the point of view of their compatibility in recycling.

Based on the results of the analysis, it may be concluded that the above-mentioned papers are expedient for the successful solution of the problem of utilization of vehicles.

However, when disposing of light commercial vehicles at the end of the life cycle, there are a number of features related to the lack of internal regulations defining the LCA design methodology, taking into account the suitability for recycling, the unpreparedness of the product composition database for calculation of recyclability indicators, the complexity of organizing work with suppliers parts and assemblies in terms of providing passports of materials and components, etc.

Depending on the choice of methodology for designing light commercial vehicles, taking into account the utilization, the output directly affects foreign markets, in particular the EU market, as well as the markets of the Russian Federation and the countries of the Customs Union, which tend to tighten environmental standards. A detailed study of the requirements of the UN Regulations and the EU Directives (UN Regulations 133, Directives 2000/53 / EC and 2005/64 / EC, ISO 22628: 2002, etc.) has shown that to bring the design of new and already produced vehicles into compliance with the requirements of recycling it is necessary to create a methodology for the design and development of light commercial vehicles that meet the utilization rates.

3. Development of a methodology for the design and development of light commercial vehicles that meet utilization rates.

At the first stages of the development of this methodology, a number of approaches were identified that should be applied to the automaker:

– Development and implementation of a set of management standards for disposal targets;

– Creation of a specialized teamwork environment for representatives of various areas of product design and production;

IOP Conf. Series: Materials Science and Engineering **386** (2018) 012017 doi:10.1088/1757-899X/386/1/012017

– Development of a design focused on the possibility of processing for the purpose of reuse;
– Early involvement of suppliers in work to ensure suitability for disposal;
– Refusal to use hazardous materials, or non-exceeding of limit values (lead, cadmium, mercury, hexavalent chromium);
– Use of materials suitable for recycling and reuse;
– Identification of components from polymers and elastomers;
– Development and provision of accessibility of Catalogs of disassembly of vehicles for recycling companies;
– Guarantee the possibility of re-use of more than 85% of the mass of the car, laid down already at the design stage.

The algorithm for designing light commercial vehicles taking into account the requirements for disposal at the end of the life cycle, combining these approaches is as follows.

At the initial design stage, the functional purpose of the product is analyzed and in parallel, according to regulatory documents, the design requirements for the product regarding the possibility of its reuse, recycling and recovery are determined. Based on this analysis, the structure of the product and the choice of its material are formed. When the product characteristics are consistent with the requirements of engineering analysis, an assessment of the suitability of the product material for its recycling is made. In the case of a negative evaluation of suitability, the possibility of changing the type of the material of the product and the ease of dismantling the product are considered. After satisfying the requirements of engineering analysis, suitability for recycling and ease of dismantling, the costs and benefits of processing the product are considered, taking into account the production budget. If the results of calculations are consistent with the specified requirements, an analysis of costs and profits in the manufacture and processing of the product is made and the start of production begins. If the results of the calculation are not in accordance with the requirements, the design of the product is changed, beginning with the first stage of this algorithm.

In the process of designing a new car, the features in terms of recycling are built into the overall work flow, presented in Figure 1.

At the product planning stage, legislative requirements are defined in the markets for the intended sale of cars. A database is formed on the composition of the product with the possibility of calculating the utilization rates. At the product development stage, the recycling requirements are included in the technical requirements for the component supplier. The design takes into account the limitations and recommendations on the suitability of vehicles for disposal, a vehicle is selected for the calculation of indicators. At the stage of preparation of production from suppliers, they receive confirmation of the compliance with the requirements for disposal in the form of certified material passports of components. In preparing the production, also, the requirements for the suitability for disposal in relation to the production process are taken into account. At all stages, the calculation of the utilization indicators and the necessary adjustment of the design and technology are monitored (in case of non-conformance detection).

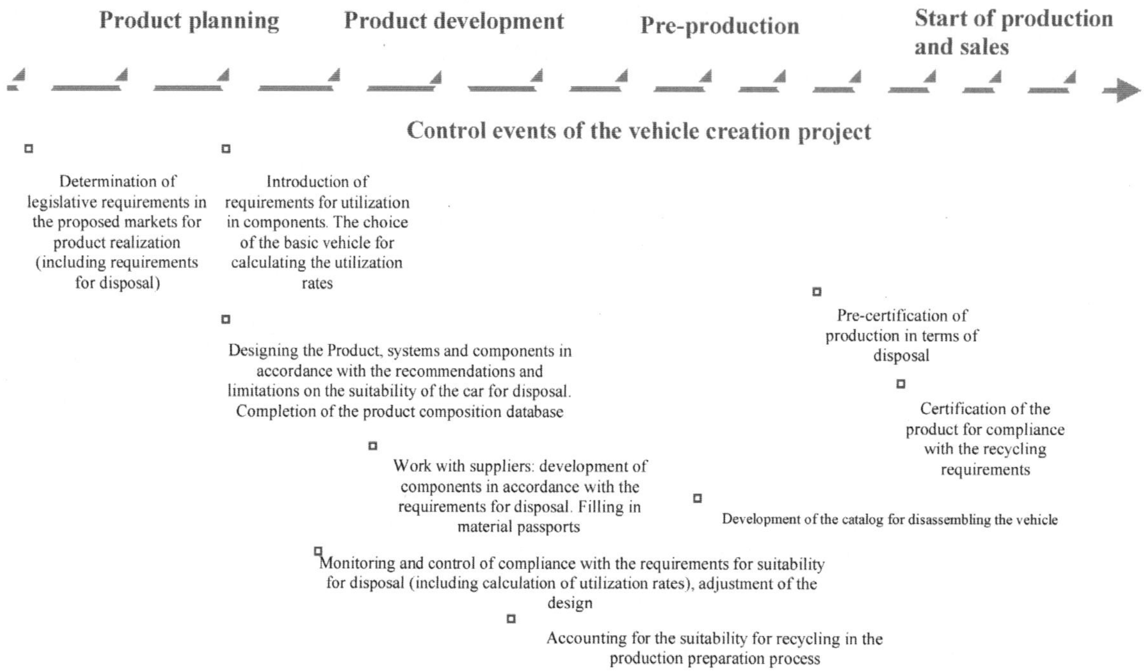

Figure 1. The work flow in order to meet the requirements for suitability for recycling when creating a product.

From the point of view of the product, the following features can be distinguished while ensuring its suitability for disposal:

Product concept:

– maximum coverage of the functional characteristics requested by the market by creating a product line with the minimum number of options required;

– exact fulfillment of customer expectations without understated or excessive product characteristics;

– Optimization of the design of the vehicle with respect to the identity of the main characteristics of components (equal resource, identical quality level);

– minimization of the material consumption and energy consumption of the product throughout the life cycle;

– design with easy disassembly. Ensuring the resistance of compounds to the effects of time (corrosion, aging, change in properties), i.e. whether it will be possible to disassemble the car in 10 ... 15 years as easily as at the time of assembly;

– minimization of the nomenclature of components from various materials, i.e. it is necessary to strive to use the same components and materials.

Use of environmentally friendly materials in the design:

– minimization of the use of hazardous materials (lead, mercury, cadmium, hexavalent chromium);

– preference for the selection of materials having the lowest energy input in the production;

– use of materials most suitable for reuse;

– reuse of materials obtained as a result of disposal;

Reducing the mass-dimensional characteristics of the product components:

– reducing the mass and volume of components contributes to reducing logistics costs in production and power consumption during the operation of the car;

Optimization of production technology:

– the use of technological processes with a minimum energy intensity and the lowest emissions of harmful substances;

IOP Conf. Series: Materials Science and Engineering **386** (2018) 012017 doi:10.1088/1757-899X/386/1/012017

– minimization of the number of technological operations;
– use of materials that do not require additional surface treatment of parts;
– minimization of waste generation;
– recycling of waste in the same process;
– the use of secondary materials in the production of cars.

Use of the product at the end of its life cycle:
– at the design stage, to consider the possibility to reuse components after appropriate repair;
– the design should allow easy disassembly and separation of materials;
– use a modular design that simplifies disassembly;
– Materials must be labeled (especially polymers and elastomers) to speed up their sorting;
– The more materials are recyclable - the less waste will be left for burial;
– parts and assemblies, which containing hazardous materials must be easily removed to avoid further contamination of the environment (especially when burning the waste).

Since the indicators of suitability for utilization and recovery are determined by the calculation of the masses of materials of vehicle components identified in the technical documentation of the automaker with subsequent monitoring of actual deviations in production - the issue of mass management is one of the key values in the management of disposal targets. The procedure for controlling the target vehicle mass is shown in Table 1

Table 1. Procedure for controlling the target vehicle mass

№	Action	Stage of product creation
1	Analyze trends in the change in the mass characteristics of cars and their components in future periods. Carry out an analysis of previous models of own production. Conduct an analysis of the mass values of products of competitors and their components in the work of benchmarking.	Planning
2	Set the target masses for the basic modifications of the car in the product family	Planning
3	To present the engineering composition of basic modifications to large systems - the first level of decomposition (body, chassis, electronics, ...). Carry out a breakdown of the target mass of the product into systems.	End of planning phase
4	Carry out the decomposition of systems into subsystems (wheels, gearbox, ...) - the second level of decomposition. Carry out a breakdown of the target mass into subsystems.	Start of development
5	Decompose the subsystems into components - the third level of decomposition. Carry out a breakdown of the target mass into components.	Development
6	Development of components design in the framework of non-exceeding of the target mass. Correction of mass distribution in justified cases.	Development
7	Confirmation of achievement of target values in production. Carrying out structural changes in justified cases.	Pre-production

Calculation of recycling indicators is carried out on the basis of a complete set of information on the composition and masses of each component of the vehicle, up to fasteners, seals, fillers, solders, etc. The relationship between the stages of recycling the vehicles and calculating the suitability for utilization is shown in Figure 2.

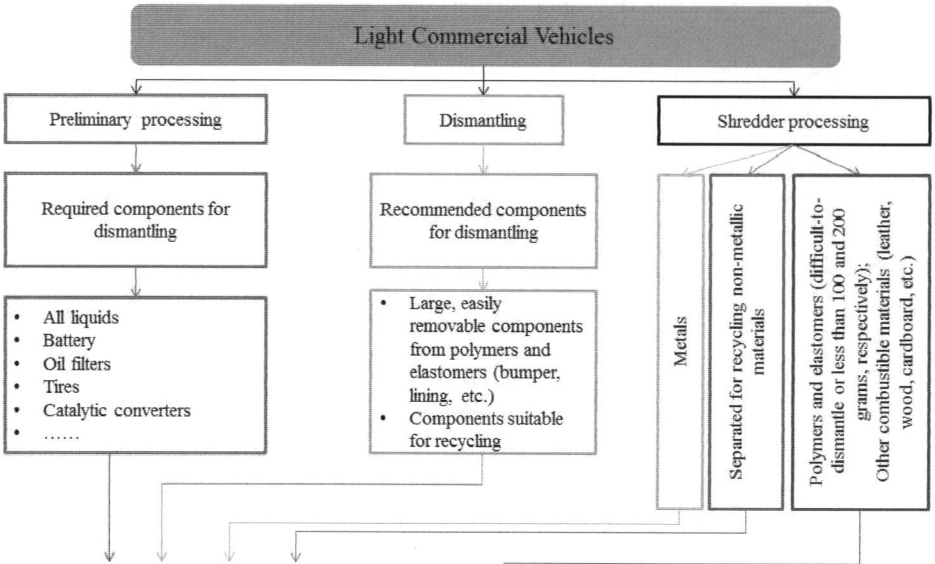

$$R_{cyc} = \frac{m_p + m_D + m_M + m_{Tr}}{m_V(curb\ weight)} \times 100 \geq 85\%$$ - recyclability indicator for reuse

$$R_{cov} = \frac{m_p + m_D + m_M + m_{Tr} + m_{Te}}{m_V(curb\ weight)} \times 100 \geq 95\%$$ - recovery indicator

Figure 2. The relationship between the stages of recycling of the vehicle with the calculation of the suitability for utilization

Another key point in the suitability of cars for utilization is the provision of an efficient process of disassembling the car at the end of the life cycle [7,8]. This criterion is directly related to the economic efficiency of car recycling. The time for disassembling the car is determined by the features of its design, the elements of fastening the units and parts, the availability of instructions for disassembling, etc. To date, many foreign automakers are paying special attention to the time costs associated with disassembling the car, as the disassembly time increases, the total cost of car recycling also increases.

In view of the foregoing, in order to minimize the time and financial costs for the disposal of light commercial vehicles, special solutions and limitations are being put at the design stage, simplifying disassembly. Using the CAD system and a specialized process simulation system, an imitation of the component is simulated with the calculation of the trajectory and the time of operations. An example of modeling the dismantling of plastic parts of the GAZelle Next cladding is shown in Figure 3.

To evaluate the effectiveness of the dismantling of GAZelle Next plastic details, which was designed to take into account the dismantling requirements for disposal, and GAZel Business (GAZ 3302), which did not have the requirements for dismantling the components, comparative tests were carried out. The essence of the tests was to estimate the number of plastic components and the time taken to dismantle them. Despite the fact that the number of plastic parts in the GAZelle Next car, compared to GAZelle Business, is more by 15%, the time spent on dismantling this type of products is less by 25%. Based on the results of this experiment, it can be concluded that this approach allows to minimize the time and material costs for ensuring the dismantling of components, as well as to increase the number of dismantled non-metallic parts, which increases the car's availability for recycling.

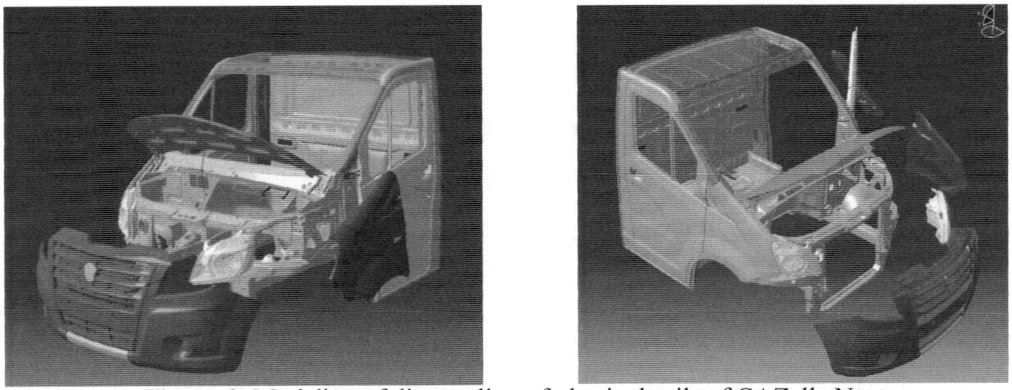

Figure 3. Modeling of dismantling of plastic details of GAZelle Next

The result of the above works is the catalogs for the disassembling of light commercial vehicles of the GAZ Group, which meet the requirements of the EU regulations on the suitability of vehicles for disposal.

4. Conclusions

Thus, to ensure the leading position of the LCA GAZ on the domestic market and the growth of sales in foreign markets, a significant increase in the environmental safety of products was achieved, including by ensuring the suitability for disposal at the end of the life cycle. The results of this work allowed to implement the project to develop new and bring already produced light commercial vehicles GAZ in line with the European requirements for the suitability for disposal at the end of the life cycle.

It is worth noting that, with the direct application of the above-described methodology for designing light commercial vehicles taking into account disposal at the end of the life cycle, certificates of compliance with Directive 2009/1 / EC and 2005/64 / EC were obtained for production and products respectively, which is one of the conditions for access to export markets.

Acknowledgments

This research done with the financial support from Ministry of Education and Science of the Russian Federation (the unique project identifier is RFMEFI57717X0268).

The experimental research conducted with the use of measurement equipment of the NNSTU Centre of collective using «Transport Systems».

References

[1] Akulova, A.A. Process Organization of vehicle recycling in the Ural region: *dis. ... cand. tech. science*: 05.02.22 / Ekaterinburg, 2017.

[2] The official website of the Russian newspaper. Retrieved May 30, 2018, from: https://rg.ru/2013/09/10/transport.html

[3] Bobovich, B. B. Utilization of vehicles and automotive components: *Textbook* / BB Bobovich. - Moscow: MGIU, 2010. - 176 p. - ISBN: 5-91134-504-8 978-5-91134-504-4

[4] Trofimenko Yu.V., Vorontsov Yu.M., Trofimenko K.Yu. Car recycling: *Scientific monograph* / M: AKPRESS, 2011, 336 p. ISBN 978-5-91293-066-9

[5] Petrov, RL Ecological safety of VAZ cars in full life cycle: *dis. ... cand. tech. sciences*: 05.05.03 / Moscow, 2006.

[6] Sychev A.V, Terenchenko A.S. Methodology of designing cars in view of their subsequent disposal / *Zurnal AAI*, №1, 2012, pp. 34-35.

[7] Kryaskov V.G., Gagunov S.A., Groshev A.M. Development of a methodology of creating ELV dismantling catalogues / *Modern problems of science and education*, №6, 2013, pp. 109-117.

[8] V.G. Kryaskov, T.V. Anfimova, S.A. Gagunov Analysis of methods of elv dismantling process for the needs of recycling / *Transactions of Nizhni Novgorod state technical university n.a. R.Y. Alexeev*, 2014. №1 (102). pp. 142-149.

IOP Conf. Series: Materials Science and Engineering **386** (2018) 012018 doi:10.1088/1757-899X/386/1/012018

Modular heavy duty truck transmission

I R Mavleev, I I Salakhov and D I Nuretdinov

Kazan Federal University, Kazan, Russian Federation

E-mail: mirkampi@mail.ru

Abstract. The design of a modular gearbox for heavy trucks, which consists of a differential divider, a main gearbox and a differential demultiplicator with an integrated rear gear is described. The proposed design has smaller dimensions, weight and inertial masses of rotating parts, has greater rigidity and speed of the gear shifting process.A vehicle equipped with this transmission will make better use of the engine power and save fuel, which will increase its competitiveness in the market.

1. Introduction

For heavy goods vehicles to increase the range of gear ratios it is necessary to use a gearbox with a combined kinematic scheme which consists of a main reducer, formed by the several pairs of gears located on parallel shafts fixed axes of rotation, and an additional gear (dividers and a demultiplicator), and a divizor which is a two-step gear drive and a dual planetary gear mechanism [1, 2].

It is obvious that designers of multi-stage gearboxes have the following tasks:

- ensuring the general requirements for any new vehicle units, namely improved reliability and rigidity, as well as modularity of the structure with smaller dimensions and weight;
- implementation of special requirements for gearboxes, namely, improving speed, providing the required number of stages and the required range of gear ratios, ensuring smooth gearshift without breaking the power flow, providing the ability to test individual modules of the gearbox, the transmission of full power when reversing, etc.

2. Analogues of the proposed transmission

The devices which are the closest in terms of the set of characteristics to the claimed invention for the step change of the torque and its transfer to the driving wheels of the car are known. For example, the most successful European version of the gearbox for trucks is the ZF Ecosplit (fig.1.) [3].

They are a mechanical gearbox, consisting of a main four-stage gearbox, a two-stage gear divider, built into the main gearbox, and a two-stage demultiplicator, formed by a three-tier planetary gear.

Arrangement of the divider consisting of two pairs of gears and a synchronizer in the main gear housing increases the length and weight of the secondary and intermediate shafts, respectively, increases the dimensions of the main gear. In addition, the main gearbox has an additional number of gears to provide a rear gear, which also increases the length of the shafts and dimensions of the main gearbox housing. A number of gears of the rear transmission, due to design features, do not provide the transmission of full engine power in case of necessity to use it in the operation of the car.

The synchronizer is used to switch the divider switches four times more often than each synchronizer of the main box when driving the car, which leads to more intensive wear. In this case, each switch has the divider off-on clutch, which leads to increased wear of friction clutch parts, the rupture of the power flow transmitted to the driving wheels of the car. This reduces the utilization of the engine power, decreases efficiency of the box and reduces the comfort of driving.

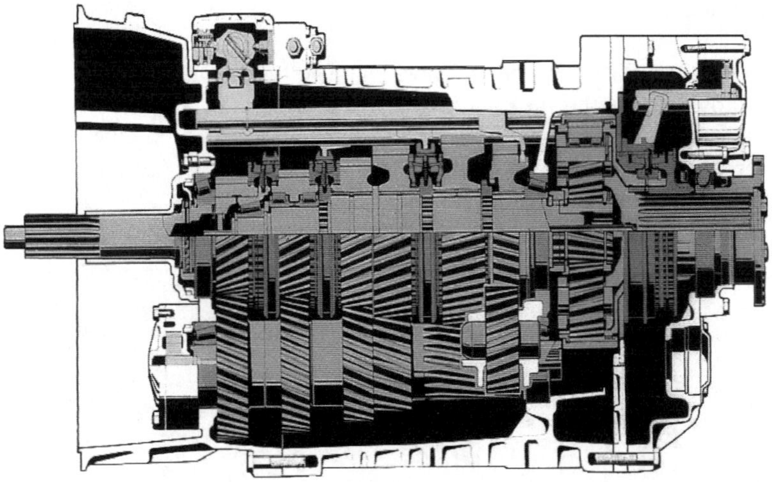

Figure 1 – The design of the multistage gearbox ZF Ecosplit

3. Modular gearbox design

A modular gearbox design, which consists of the following modules: a main gearbox, a differential divider and a differential demultiplicator, was made in the Naberezhnye Chelny Institute (branch) of the Kazan Federal University.

3.1. *Main gearbox reducer*.

The kinematic scheme and the design of the main gearbox are shown in fig. 2.

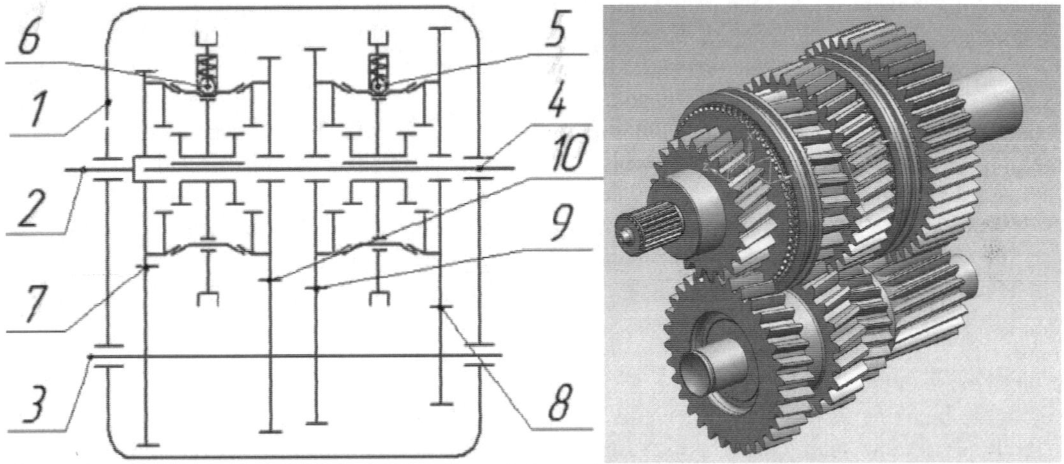

Figure 2 – The kinematic scheme and design of the main gearbox: 1 – a main gear housing; 2 – a main shaft of the main gearbox; 3 an intermediate shaft of the main gearbox; 4 – a secondary shaft of the main gearbox; 5 and 6 – synchronizers, respectively, the first-second and third-fourth gear of the main gear; 7 – a pair of permanent engagement of the main gear; 8 – a pair of the first gear of the main gear; 9 – a pair of the second gear of the main gear; 10 – a pair of the third gear of the main gear.

The main gear housing 1 with the clutch housing forms a closed cavity in which main shaft 2, intermediate shaft 3 and secondary shaft 4 of the main gearbox are located. The input shaft is the shaft-gear resting on the roller bearing, installed in the hole of the clutch housing. The intermediate shaft is supported by two roller bearings installed in the housing and clutch housing openings. The secondary shaft is supported by a roller bearing mounted in the main shaft with one end and a roller bearing mounted in the main gear housing with the other end.

The main gearbox reducer has four gears, which are switched by means of synchronizers 5 and 6. Synchronizer 5 includes the first and second gear of the main gear, while the synchronizer 6 is in the neutral position. The first gear of the main gear is activated by moving synchronizer 5 clutch to the far right position when the first gear wheel with the secondary shaft 4 is locked. The second gear of the main gear is activated by moving the clutch synchronizer 5 to the far left position when the second gear is locked with secondary shaft 4. The third and fourth gears are switched on by the synchronizer 6, while the synchronizer 5 is in the neutral position. The third gear is activated by moving the main gear clutch synchronizer 6 to the far right position when the third gear is locked with the secondary shaft 4.he fourth gear of the main gear is direct, it is activated by moving the synchronizer coupling 6 to the extreme left position when the main shaft 2 is blocked with the secondary shaft 4.

3.2. *Differential divider.*

The kinematic scheme and the design of the main gearbox are shown in fig. 3.

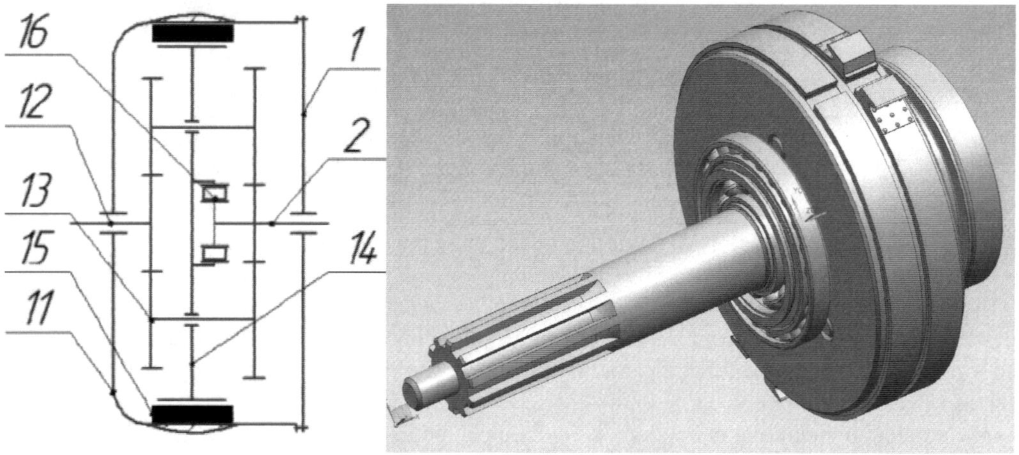

Figure 3 – The kinematic scheme and the design of the differential divider: 1 – a main gear housing; 2 – a main shaft of the main gearbox; 11 – a clutch housing (crankcase differential divider); 12 –an input shaft-gear differential divider; 13 – two-crown satellites of the differential divider; 14 – a carrier differential divider; 15 –a band brake; 16 – a free-wheeling clutch

The differential divider contains an input shaft-gear 12, which is engaged with two- crown satellites 13 mounted on the axes pressed into the carrier of the differential divider 14. The carrier of the differential divider is supported by ball bearings mounted in the clutch housing and in the differential divider cover, and is locked by the band brakes 15. Roller free-wheeling clutch 16 is installed between the carrier 14 and the main shaft of the main gearbox 2.

The differential divider has two gears. The accelerating transfer of the divider is activated by blocking the carrier of the divider 14 on the clutch housing 11, which are carried out by the band brakes 15. The direct gear of the divider is switched on automatically when the band brakes are released by locking the divider to the free-wheeling clutch 16 mounted on the main gear shaft 2. Automatic blocking of the divider drive to the free-wheeling clutch is possible due to the features of differential mechanisms with a positive gear ratio less than one, in which all the links of the differential mechanism rotate in one direction, and the carrier has greater angular rotation speed than angular velocities of the input and output shafts. Switching the range of the differential divider is carried out without switching off the clutch and, therefore, without breaking the power flow. The brake torque on the band brakes is a part of the motor torque transmitted to the output shaft. For example, when the gear ratio of the differential divider i equals 0.75, the band brake operates 0.25 torque and the output shaft 0.75 torque developed by the engine. The hydraulic system band brake switching algorithm ensures fast and smooth switching of the differential divider ranges.

A small difference in the gear ratios of the neighboring gears of the gearbox allows you to choose the optimal driving mode in the economic range of the engine speed. In addition, it facilitates the control of the gearbox and reduces the noise level.

In contrast to the prototype, in which the switching range of the divider is carried out by means of a synchronizer, in the proposed gearbox to enable the accelerating transfer is braking the carrier of the differential divider by the band brakes. To enable a direct gear, the band brakes are released, and the differential divider is automatically locked by the free-wheeling clutch onto the output shaft of the differential divider.

3.3. *Differential demultiplicator with an integrated reverse gear.*

The kinematic scheme and construction of the differential demultiplicator are shown in fig. 4.

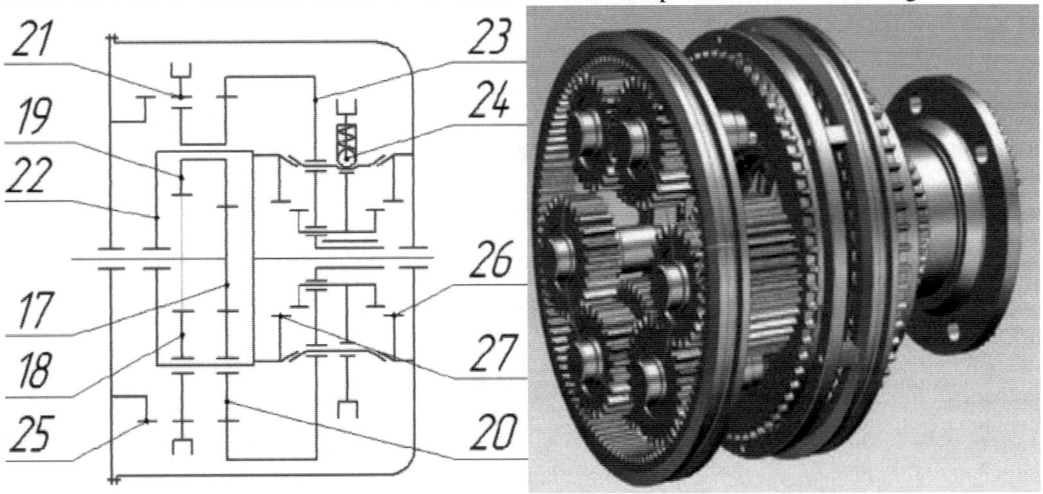

Figure 4 – The kinematic scheme and the design of the differential demultiplicator: 17 – a leading gear ring of the differential demultiplicator; 18 – satellites of the differential demultiplicator; 19 – two-crown satellites of the differential demultiplicator; 20 – reverse idler satellites; 21 – a crown wheel (reverse clutch) of the differential demultiplicator; 22 – a carrier (output shaft of the transmission) of the differential demultiplicator; 23 – a crown wheel; 24 – a synchronizer of the differential demultiplicator, 25 – a ring gear lock of the reverse clutch; 26 – a toothed crown locking the crown wheel of the differential demultiplicator; 27 – a ring gear locking differential demultiplicator

The kinematic scheme and construction of the differential demultiplier are shown in Fig. 4. The differential demultiplicator comprises a drive gear ring 17, cut on the secondary shaft of the main gear, satellites 18 and twin satellites 19, parasitic satellites 20, a crown wheel 21, which is simultaneously a reverse clutch, a demultiplier 22, which at the same time output shaft of the gearbox, and a crown wheel 23. There is a synchronizer for dual 24 on the splines of the hub of the crown wheel 23.

The differential demultiplicator with integrated reverse gear has two forward and neutral gears, one reverse gear. When driving a loaded truck in heavy traffic conditions included a reduction gearing of the differential demultiplicator. For this purpose, the synchronizer 24 gear coupling moves to the extreme right position and, after aligning the angular velocities of the connected parts engages with the toothed crown 26. When switching the differential demultiplicator to the upshift, the gear coupling of the synchronizer 24 moves to the extreme left position and engages with the gear ring 27. The reverse gear is switched on by moving the crown wheel 21, which is also the reverse gear, to the far left position and engaging it with the gear ring 25, while the synchronizer clutch 24 is moved to the neutral position.

4. Conclusions

The developed modular transmission for heavy trucks provides the following technical results:

1. The exclusion of the old divider design, consisting of two pairs of gears and a synchronizer, in the main gearbox housing, with its replacement by a differential divider, which provides reduction in the axial dimensions of the gearbox, the weight of the secondary and intermediate shafts.

2. The differential divider module also provides smooth gear ratio control, which allows to switching the divider range without turning off the clutch, smoothly and without breaking the power flow. This provides some reduction in wear of friction clutch parts, increased power utilization and increased fuel efficiency of the engine, as at the moments of switching the range of the divider, which occur four times more often than the gear shift of the main gearbox, the engine does not go into partial load modes.

3. The differential divider is controlled by a hydraulic drive, which is activated only by means of a button located on the shift lever, which facilitates control when driving, increases comfort and reduces the physical load on the driver.

4. Elimination of extra rows of toothed gears reverse through the use of the differential demultiplicator with integrated reverse gear. This allows you to get multiple reverse gears, and also to

transfer the full power of the engine when it is reversing at different speeds, which is especially important for special purpose vehicles.

5. The differential demultiplicator with integrated reverse gear further reduces axial dimensions and weight of the secondary and intermediate shafts, which increases the rigidity of the main gear shafts and their moments of inertia. This contributes to a faster alignment of the angular velocities of the rotating masses, reducing the load on the synchronizers, and therefore, increases their durability.

6. Increasing the reliability of the gearbox, as the technological process of its assembly can be carried out after the preliminary bench tests of its individual components, namely, of the differential divider, of the main gearbox and of the differential demultiplicator, which in its turn creates prerequisites for automating the process of the gearbox general assembly.

Thus, the developed modular transmission meets all the requirements for modern units of heavy trucks.

References

[1] Faskhiev, H.A. New transmission for vehicles / H.A. Faskhiev, V.V. Voloshko, I.R. Mavleev, I.I. Salakhov // Handbook. Engineering journal with Appendix. – 2015. – №7. – p.47–51.

[2] Patent. 2508486 RU. Automotive multi-stage transmission / V.V. Voloshko, I.R. Mavleev, I.I. Salakhov. – Application № 2012148057/11; application date 12.11.2012; publishing date 27.02.2014.

[3] Hassle, H. 'Ecosplit' – a new gearbox series for heavy commercial vehicles / H. Hassle // AUTOMOBILTECH. Z. 82(7-8, Jul.-Aug. 1980), c. 369-374.

[4] Blokhin, A. Research of robotized manual transmissions for all-terrain vehicles / A. Blokhin, L. Barakhtanov, E. Fadeev, P. Lubichev // ARPN Journal of Engineering and Applied Sciences, 2017, 12 (1), pp.20-32.

[5] Salaкhov, I.I. Kinematic scheme and design of automatic planetary gear boxes based on a new module / I.I Salaкhov, V.V. Voloshko, I.D. Galimyanov, I.R. Mavleev. // Contemporary Engineering Sciences, Vol. 8, 2015, No.1, pp.1-6.

[6] Salaкhov, I.I. Car gearbox on the basis of the differential mechanism / I.I. Salaкhov, I.R. Mavleev, E.N. Tsybunov, R.R. Basyrov and N.I.Salakhov // Biosciences Biotechnology Research Asia, 2015; Vol. 12(Spl. Edn. 2), 41-44 pp.

[7] Naunheimer, H. Automotive transmissions: Fundamentals, selection, design and application / H. Naunheimer, B. Bertsche, J. Ryborz, W. Novak, P. Fietkau // Automotive Transmissions: Fundamentals, Selection, Design and Application. pp. 717

[8] Salaкhov, I.I. Development of a gear box of the truck / I.I. Salaкhov, I.R. Mavleev, I.R. Shamsutdinov, D.I. NuretdinovE.N., and N.I.Salakhov // Biosciences Biotechnology Research Asia, 2016; Vol. 13(2), 859-864 pp.

[9] Salaкhov, I.I. Analysis workflows gear hydraulic machines/ I.I. Salaкhov, I.R. Mavleev, V.V. Voloshko, I.D. Galimyanov, R.K. Takhaviev // Biosciences Biotechnology Research Asia, 2016; Vol. 13(2), 779-784 pp.

[10] Liao, Y.-G. Experiment and simulation of medium-duty tactical truck for fuel economy improvement / Y.-G. Liao, A.M. Quail Jr. // Energies, 4 (2), 2011, pp. 276-293. http://www.mdpi.com/1996-1073/4/2/276/pdf doi: 10.3390/en4020276

[11] Blokhin, A. Multistage mechanical transmissions with automatic control for advanced trucks and buses / A. Blokhin, A. Nedyalkov, L. Barakhtanov, A. Taratorkin, A. Kropp // Acta Mechanica et Automatica, 11(4), 260-266 pp.

[12] Karpukhin, K.E. Comprehensive life cycle analysis of different types of energy storage for electric or hybrid vehicles / K.E. Karpukhin, A.V. Kozlov, S.V Bakhmutov, A.S. Terenchenko // ICAT 2015 Proceeding the International Conference on Automobile Technology for Vietnam, 2015 pp. 7-14.

[13] Mavleev, I.R. Development of efficient schemes and designs high-torque hydromechanical CVTs for vehicles: Author. dis. Cand. tehn. Sciences. - Naberezhnye Chelny, 2007; 19.

IOP Conf. Series: Materials Science and Engineering **386** (2018) 012019 doi:10.1088/1757-899X/386/1/012019

New testing methods of automatic emergency braking systems and the experience of their application

A M Ivanov, S R Kristalniy, N V Popov, M A Toporkov, M I Isakova

Moscow Automobile and Road Construction State Technical University (MADI), Moscow

ivanov-am@madi.ru, sportauto@rambler.ru, niko-popov@yandex.ru, makstoporkov@rambler.ru, marina_sy@mail.ru

Abstract. This article offers testing methods of automatic emergency braking systems ("AEBS") for automobiles. Checking of the "AEBS" is carried out during the vehicle's moving towards different static "targets". It is estimated a possibility of the "AEBS" to recognize different "targets", the warning moment of collision situation and automatic braking effectiveness. The test drives are performed on different support surfaces. To avoid the damage of an automobile during the crash with the "target" or the "soft target" is used. This target is made of cardboard material. Besides, avoidance of crashing may be achieved by performing the maneuver "detour of obstacle". The method of synchronization of a collision warning signal with a parametric record of arrival is described also. There were performed tests of the "AEBS" of "Infiniti QX 60 Hi-tech" automobile model. Also there we analyzed reasons of the "AEBS" failures in various conditions. It was estimated that the technical vision units of the "AEBS of "Infiniti QX 60 Hi-tech" automobile model recognize the "soft target" as the obstacle.

Introduction

At present time a lot of car manufacturers offer driver assistance systems (DAS/ADAS) as an option on their automobiles, among them are: lane keeping systems, automatic emergency braking systems (AEBS), adaptive cruise control systems, automatic parking systems and many others. These systems, despite the additional cost, are very popular among customers. Thus, according to "Auto revue" magazine [1], a 15% increase of Subaru brand customers' amount in Europe was provided not by the all-wheel drive option or the boxer engine of these automobiles, but by the "EyeSight" brand security system. In turn, the customers expect safety of "DAS/ADAS" during the drive mode, which should be checked during the tests. Article [2] describes the possible problems during the operation of vehicles with similar systems and indicates the necessity for their testing.

As a part of the scientific work of the "Automobiles" chair of MADI, performed according to the State task of the Ministry of Education and Science of Russian Federation, it was decided to check the effectiveness of the "AEBS" of an automobile during various road situations. Test procedures were carried out to obtain information about the operation of a typical modern "AEBS".

The aims of the research works were:

1. Definition of terms and conditions (the speed of object, distance to the "target", type and offset of the "target") actuation or failure conditions of the "AEBS";

2. Determination of failure statistics of the "AEBS";
3. Evaluation of the moment in time receipt of the collision warning signal;
4. Evaluation of automobile's deceleration value, performed by the "AEBS";
5. Analysis of the "AEBS" adaptability due to changes of road surface conditions;
6. Determination of the "AEBS" resistance to false actuations.

Under the "AEBS Failure" it was considered the situation of non-response collision warning signal.

To perform experimental studies, a test program and a method of performing of run tests were developed, as well as a measuring complex was assembled to register the parameters of the test object (TO).

Test program and test procedure

To check the main functions of the "AEBS", which determine the overall efficiency of the system, various road situations were modeled in accordance to table1.

Table 1 Test program AEBS

Type test	Man oeuvre	«Objective»	Test object speed, V, км/h	«objective» displacement Elat, %	Road surface
1	Passing obstacle	Soft wall	40 60 80	0	Dry asphalt
2a	Braking; Steadiness on a straight course (driver brakes)	Soft wall	40 60 80	0	Dry asphalt Wet asphalt Wet basalt
2б	Braking; Steadiness on a straight course (AEBS brakes)	Soft wall	40 60 80	0	Dry asphalt Wet asphalt
3	Passing obstacle	Soft wall	40 60 80	50	Dry asphalt
4	Passing obstacle	Vehicle.	40 60 80	0	Dry asphalt
5	Passing obstacle	Motorbike model	40 60 80	0	Dry asphalt
6	Passing obstacle	Bicycle	40 60 80	.0	Dry asphalt
7	Passing obstacle	Dummy	40 60 80	0	Dry asphalt
8	Birec troute	Vehicle	40 60 80	100	Dry asphalt

Test runs were carried out on dry and wet asphalt road, as well as on wet basalt road (for some types of test procedures), simulating a coating with a low adhesion coefficient.

Test types1-7 allow to define resources of "technical vision" of the "AEBS" detection methods of various types of static obstacles and the ability of the "AEBS" to reduce the speed automatically. When performing tests1-7, the vehicle moves in a straight line towards the "target".

IOP Conf. Series: Materials Science and Engineering **386** (2018) 012019 doi:10.1088/1757-899X/386/1/012019

In tests1, 3-7 after receiving a collision warning signal, the driver does not take measures to reduce the speed. Regardless the fact of operation of the "AEBS" during test procedures 1, 3-7, the test driver performs a maneuver of the bypass of the "target" to avoid a collision with it (Figure 1). This maneuver is performed with minimum safe distance from the "target". The maneuver of obstacles' detour allows to avoid a contact with the "target" much more exactly and it may be carried out later than the maneuver "braking in front of an obstacle" and allows to analyze the work of the "AEBS" almost to the obstacle itself.

Figure 1. The maneuver around an obstacle (soft wall)

During the test2a after the receipt of a collision warning signal, the test driver performs an emergency braking. The object during the slowdown remains in the lane until its stop in front of the "target" or before hitting the "target".

Test2b is identical to test2a and differs in a matter that after the receipt of collision warning signal the test driver does not perform braking and expects the object to slow down from the action of "AEBS".Test2a allows to estimate the timeliness of the collision warning signal.

Test2b allows to determine the slowdown value, performing by the "AEBS" of test object without driver participation.

Test8 allows to evaluate the stability operation of the "AEBS" to its false actuation are guarding a static vehicle.

During the test8 the vehicle moves rectilinearly in the direction of the "target" that is installed in the adjacent lane.

There were analyzed regulations during the development of the test program [3, 4, 5, 6]. Also the experience of other similar studies has been used [7, 8, 9].

Test runs were carried out on special roads of the NAMI's Testing Centre (located in Dmitrov city).

The description of "targets":

As "targets" during the "AEBS" test procedures there were applied:

- "soft wall" (Figure 1);
- vehicle;
- motorbike model(Figure 2a);
- bicycle (Figure 2b);
- dummy (Figure 2c).

IOP Conf. Series: Materials Science and Engineering **386** (2018) 012019 doi:10.1088/1757-899X/386/1/012019

a b c

Figure 2. "Targets": a) motorbike model; b) bicycle; c) dummy

For the test runs with a probability of crashing situation of the "TO" with the "target" is used a "soft wall" made of a cardboard material. On the surface of the "soft wall" facing the approaching "TO" it is fixed a metalized coating necessary to improve the sensitivity of the "target" position. The main dimensions of the "target" are shown in figure 3.

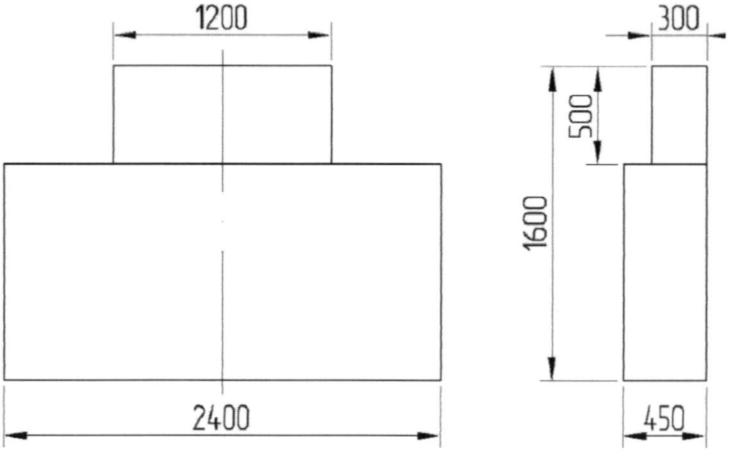

Figure 3. Main dimension features of the "soft wall target"

The main reason of the application of "soft target" is the safety improvement of test runs with the mandatory condition of adequate detection of such a "target" by the "AEBS" of tested object.

There was made a marking of test area and fixing of the geographical coordinates of the "goal" before the run tests.

Under the lateral displacement of the target Elat it is concerned the transverse distance between the longitudinal axes of the test object and the "target" measured as a percentage of the width of the "TO" (Figure 4). At Elat=100 % the "target" is located in the adjacent lane with the maximum offset from the side surface of the "TO" not more than 0.3 m along the outer rear-view mirrors.

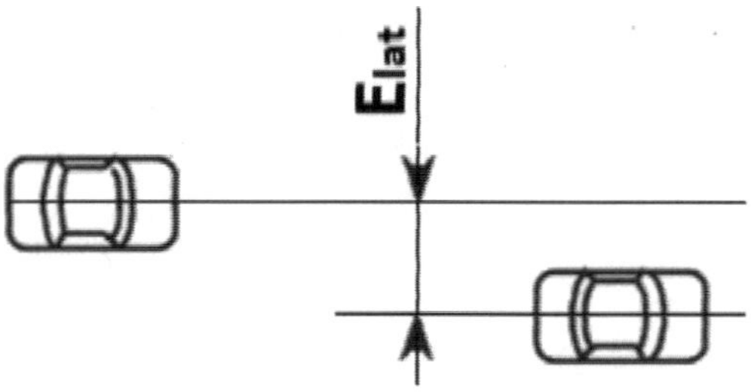

Figure 4. Elat lateral displacement between the "TO" and the "target"

Fixing method of the operation moment of the collision warning signal

To determine the parameters of the vehicle movement during the visual and/or acoustic warning signal of the collision situation it has been developed a method that allows to synchronize the time of the signal with the parametric record of the test run. The advantage of the proposed method is the ability to record any audio-visual signals regardless of origin: the appearance of an audio signal, the light control lamp on the instrument panel, the message on the screen of the on-Board computer, etc. Disadvantages of this method are the inability to fix the tactile signals (increased force on the accelerator pedal, belt tension, rapid failure of the brake pedal, etc.), the complexity and time of data processing. The fact of providing of the collision warning signal is fixed by the video camera. In the "field of view" of the camera there is a lamp (LED) of internal light indication, which lights up in a certain time after the start of parametric recording. In the processing and recording unit it also records the moment of filing and the duration of the digital signal, which controls the activation of light indication lamps. During the record procedure made by the camera, you can synchronize the time of the receipt of the collision warning signal with the parametric recording. The fact of presence of an external indicator lamp allows, in the presence of an external video, to synchronize with the parametric record all processes that occur with the automobile during the test run, for example, the moment of switching on the brake lights.

Test object ("TO")

It was submitted "Infiniti QX 60 Hi-tech" automobile model by "INFINITI Russia" Company for testing procedure. This automobile is equipped with the "FEB" system (Forward Emergency Braking), an emergency braking system with the detection option of pedestrians, is used in case of a frontal collision danger [10].

The "FEB" system uses radar (Figure5, POS. (a)) installed behind the lower grille of the front bumper to measure the distance to the vehicle moving in the same lane. A camera is provided to detect pedestrians (Figure5, Pos. b) [10]. This camera is installed on the top behind the windshield. The "FEB" system operates at speeds of more than 5 km/h (in the speed range from 10 km/h to 60 km/h when pedestrians are detected).

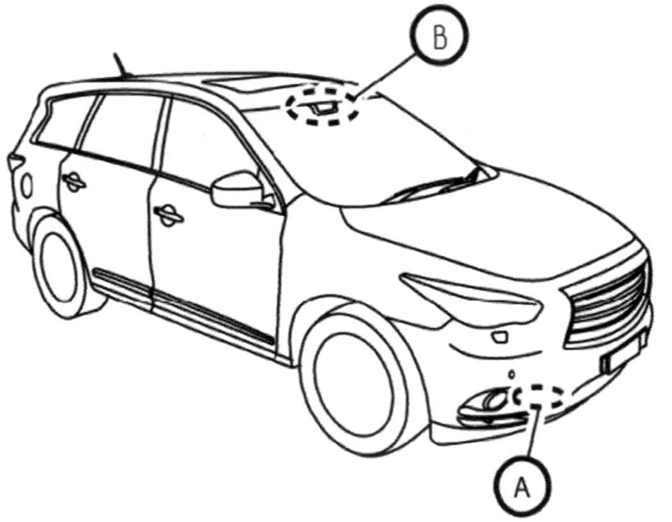

Figure 5. Means of "technical vision" of the driver assistance system's units: A-radar, B-camera [10]

Measuring and recording equipment

To solve the above mentioned problems of research work it was installed the following measuring and recording equipment on the automobile:

- torque measurement steering wheel MEASUREMENT STEERING WHEEL (MSW) produced by "KISTLER Company", Germany;
- brake pedal force sensor "CPFTA" produced by "KISTLER Company", Germany;
- remote angular velocity wheel sensor "WPT" produced by "KISTLER Company", Germany determines the circumferential speed of the left front wheel;
- acceleration and angular velocity sensor "Tri-Axial Navigational Sensor" (TANS) produced by "KISTLER Company", Germany;
- GPS antenna produced by "IMC" (Germany) and "JAVAD" (USA) Companies;
- universal measuring system for data collection and processing "CS 1016 FAMOS Online" produced by "IMC Company", Germany.

The power for the equipment was provided from an onboard power supply network of the vehicle via the power distribution unit "Small 12VPowerDistributorBox" produced by "KISTLER Company", Germany.

Registration of visual warning signal of the collision on the dashboard was performed by video camera mounted on the steering column cover.

Processing of test results

Time dependence diagrams were submitted for each test run (Fig. 6-8):

- movement speeds V;
- the longitudinal deceleration (jx);
- the angle of the steering wheel (α);
- distances from the "TO" to the " target "(D);
- the value of push effort on the brake pedal (Pbped).

The start point on the diagrams is concerned the warning moment of a possible collision situation. The distance between the "TO" and the "target" is determined by the coordinates of the global positioning system.

Examples of the diagrams for different types of tests are shown on figure 6.

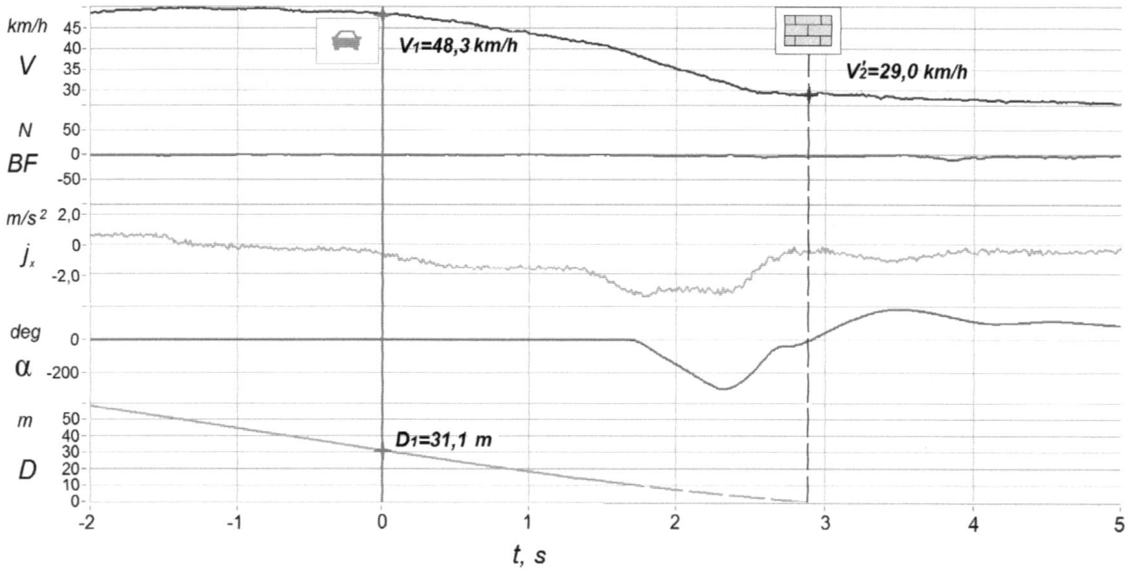

Figure 6a. Example of the diagram for tests 1, 3, 4, 5, 6, 7, 8

IOP Conf. Series: Materials Science and Engineering **386** (2018) 012019 doi:10.1088/1757-899X/386/1/012019

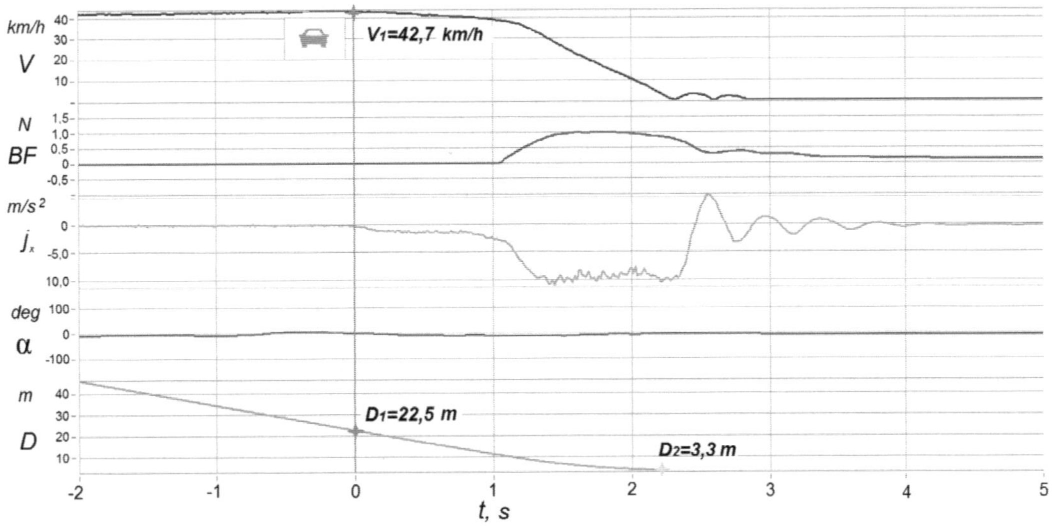

Figure 6b. Example of the diagram for test 2a

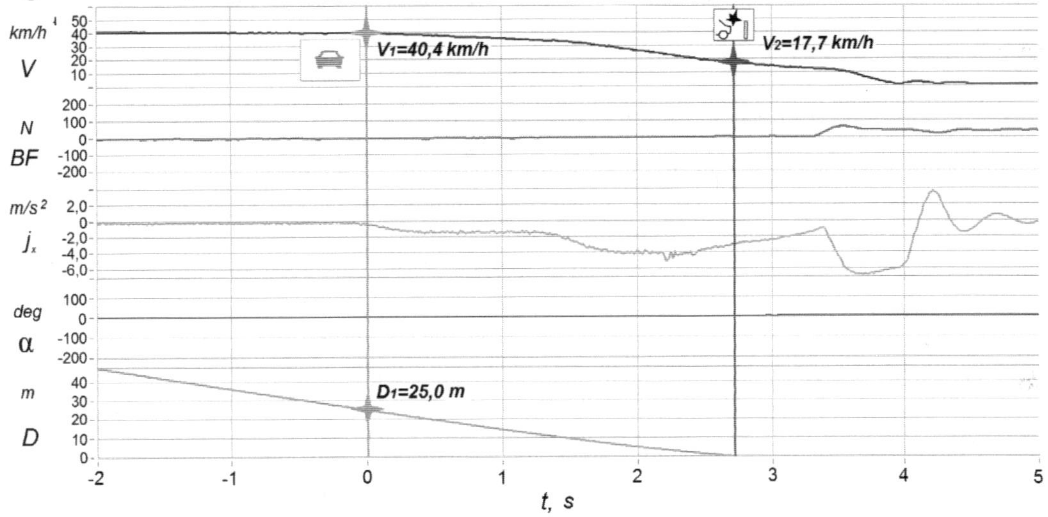

Figure 6c. Example of the diagram for test2B

Symbols on figure 6:

V1 is the longitudinal velocity of the "TO" during the warning signal moment of possible collision situation; D1 is the distance between the "TO" and the "target" during the warning signal moment of possible collision situation; D2 is the distance between the "TO" and the "target" in the case of its stop in front of the "target"; V2 is a longitudinal velocity of the "TO" during the collision with the "target"; V2 is hypothetical longitudinal velocity of the "TO" at during the collision with the "target" (when driving beside the "target") is the velocity of the "TO" during the collision with the "target", if there was no swerving;

-  – the moment of the collision warning signal;
- 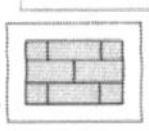 – the moment of suspended touch (when performing a detour of obstacles) of the "target" by the "TO";
-  – the moment of touch (only for soft wall) of the "target" by the "TO".

Figure 6. Example of diagrams for different types of tests

IOP Conf. Series: Materials Science and Engineering **386** (2018) 012019 doi:10.1088/1757-899X/386/1/012019

Analysis of test results

After the receipt of the collision warning signal an automatic braking action begins immediately by the "AEBS". Thus it is provided a constant value of deceleration on dry asphalt before the soft wall "targets" or the automobile at the rate 2-3 m/s ^{2} (at the speed rate from 40 km/h) and about 3-4 m/s ^{2} (at the speed rate from 60 km/h). But this efficiency of automatic braking action is not enough to stop in front of the "target".

During the test procedure 2a that was performed on dry bituminous concrete the "TO" had time to stop in front of the "target", but the collision had occurred on wet asphalt and basalt road. Thus, the "AEBS" system does not adapt to changes of adhesion coefficient value of support surface.

There were no false actions during the test procedure 8.

Failure statistics of the "AEBS" depending on different conditions is presented in table 2 and 3.

Table 2 System failures depending on the type of test

Type test	**«Objective»**	**Number of tests**		**Number of system failures (%)**	
1	Soft wall	34	Total: 59	2	Total: 4 (7 %)
2a, 2б		22		2	
3		3		0	
4	Vehicle	13		2 (15 %)	
5	Motorbike model	10		3 (30 %)	
6	Bicycle	9		3 (33 %)	
7	Dummy	18		5 (28 %)	

Table3 System failures depending on object speed

«Objective»	**System failures depending on object speed, км/h**		
	40	60	80
Soft wall	1	1	2
Vehicle	0	0	2
Motorbike model	1	0	2
Bicycle	0	0	3
Dummy	1	1	3
Total failures :	3	2	12
Total tests:	40	33	36

The analysis of table 2 shows that this "AEBS" system recognizes soft wall and the automobile "targets" in the best way.

It was also estimated the increase of the "AEBS" failures when the "TO" speed increased (see table 3).

Advance warning signal is estimated by the parameter "TTC" (Time to collision) - the time to the collision. Time to the collision is a time feature after which a collision of the "TO" and the "target" will occur as long as the current relative velocity of the "TO" and the "target" is constant.

The time to impact a stationary "target" is calculated by the following formula: TCS=D1/V1.

Table 4 shows a comparison of the average "TTC" values for different "targets" at speed rates of collision warning signal of 40 and 60 km/h.

Table 5 compares the average "TTC" values for different support surfaces.

Table 4 The TTC value for various "«Objectives»"

«Objective»	TTC, c	
	V_{cpa6}=40 км/h	V_{cpa6}=60 км/h
Soft wall	2,18	2,42
Vehicle	2,06	2,55
Bicycle	0,98	1,51

Table 5 The TTC value for various road surfaces (test 2a, 2б, «objective» – Soft wall, V=60 км/h)

Road surface	TTC, c
Dry asphalt	2,25
Wet asphalt	2,29

The proper response of the "AEBS" to a "soft target" is confirmed by a small percentage of failures (see table 2) and the identity of the "TTC" parameter for the "soft target" and the real automobile (see table 4).

The warning signal of a collision with a bicycle is received much later than with a soft wall or with an automobile. Therefore, automatic deceleration does not have time to start before the detour of obstacles.

The analysis of table5 confirms the previously made conclusion about the lack of adaptability of the "AEBS" algorithm to the change of the support surface state.

Conclusion

A new method is proposed to test automatic emergency braking systems. Features of the proposed test method are:

- tests are carried out on support surfaces with different grip value;
- investigation of the interaction with a large number of "targets";
- introduced maneuver "detour of obstacles";
- analyzed a larger number of motion parameters of the vehicle;
- new methods of processing test data are used.

The results of experimental studies of the "AEBS" of "Infiniti QX 60 Hi-tech" automobile model give us a possibility to make such conclusions:

1. Means of technical vision of the "AEBS" object recognize the type of the "target" – the "soft wall" as an obstacle. This "target" can be applied during future tests.

2. Means of technical vision of "AEBS" object recognize badly object type "target" – a "stationary bicycle". This aspect restricts the range of functional safety of the "AEBS".

3. The deceleration achieved by the object during the operation of the "AEBS" (without the driver) is not enough for the complete stop of the object in front of a static obstacle, which also reduces the effective range value of the system.

4. The collision warning signal algorithm does not depend on grip properties of the road surface. This fact reduces the functional safety of the "AEBS" on slippery surfaces.

5. A large number of the "AEBS" failures at a speed of 80 km/h indirectly indicate an insufficient distance value to detect the obstacle. This matter also reduces the range of application of the "AEBS".

6. There was not recorded an "AEBS" "false target" action.

Gratitude

The authors of this article would like to thank for the help during organization and research works:

- "INFINITI Russia" technical center in a person of its Chief Engineer Mr. Philip K. Dyakov;

- NAMI's Testing Centrein a person of Department Head of "Safety of Automobiles Department" Mr. Alexander A Barashkov and an assistant Mr Yury P Shevelkin;
- "Sensorika-M", LLC Company in a person of its Technical Director Mr. Alexander V. Zelikovskiy.

References

[1] Rastegaev O 2018 Outback in snow Autorewiev . № 6 (630) 36-37

[2] Prihodko V, Ivanov A, Borisevich V, Shadrin S 2017 Techniques for testing automated control systems Newsletter Moscow-State-Automobile-Road-Technical-University-MADI № 4(51) 10-15

[3] 2018 UNECE Regulation No. 131 (01). Uniform provisions concerning the approval of motor vehicles with regard to automatic emergency braking systems. (AEBS) - URL: https://www.unece.org/fileadmin/DAM/trans/main/wp29/wp29regs/2017/R131r1am1r.pdf

[4] Moscow: Standard inform 2017 GOST R ISO 22839-2017 Intelligent transport systems. Systems to reduce the severity of the consequences of a collision with a vehicle in front. Work, performance and verification requirements (ISO 22839: 2013, IDT). 29

[5] Moscow: Standard inform,2017. GOST R ISO 15623-2017 Intelligent transport systems. Collision avoidance systems with a vehicle in front. Performance requirements and test methods (ISO 15623: 2013, IDT). 28

[6] 2018 Safety Assist. – URL: https://www.euroncap.com/ru/для-инженеров/protocols/safety-assist/

[7] Sachkov M 2018 Top-Stop Za Rulem № 2 42-43

[8] 2016 OCAS-DRI-CIB-16-08 New car assessment program. Crash imminent braking system confirmation test. 2016 Nissan Rogue. Final Report. U. S. Department of transportation108.

[9] 2017 CAS-DRI-CIB-17-06 New car assessment program. Crash imminent braking system confirmation test. 2017 Mitsubishi Outlander. Final Report. U. S. Department of transportation, 120.

[10] 2017 Manual for infinity QX 60. - URL: https://www.infiniti.ru/content/dam/Infiniti/Russia/manuals/qx60.pdf

IOP Conf. Series: Materials Science and Engineering **386** (2018) 012020 doi:10.1088/1757-899X/386/1/012020

Overcoming of high escarp by an all-wheel drive machine 8x8

V I Peskov

Nizhny Novgorod State Technical University n.a. R.E. Alekseev, 24 Minin Street, Nizhny Novgorod, Russia

Abstract. In our research there has already been expressed the idea that from the point of view of overcoming big obstacles like escarpments, multi-axial all-wheel vehicles should have advantages over biaxial vehicles with 4x4 wheel arrangement. A thorough analysis of the movement process of a 8x8 wheeled vehicle on the escarpment which height exceeds the radius of the wheel is made. It was proved that such a vehicle can overcome the escarpment that almost twofold exceeds height of similar obstacles that a 4x4 wheeled vehicle overcomes.

Multi-axial vehicles with more than four driving wheels have the improved potential of overcoming big obstacles. For example, if an all-wheel vehicle drive with wheel formula 4x4 has a possibility to overcome the escarp by all axles, the height of which is equivalent of 0,8 wheel radius, in that case, as preliminary calculations showed, an all-wheel drive machine with a wheel formula 8x8 can overcome such an obstacle by all axles in height even more than a wheel radius [1]. It is necessary to underline only, that essential importance on these possibilities must influence the pull up of the wheels of one of the driving axles, which is inevitable in such conditions during the machine movement with a very rigid body. The matter is if the wheels of one of the axles are pulled up, then the vertical load is redistributed on other driving wheels, influencing their possibility to overcome a big obstacle seriously.

An attempt is made to study overcoming the escarp in detail, with the height being more than a wheel radius. A special armoured vehicle with an 8x8 wheel formula created by *General Dynamic* in Great Britain was used as a prototype. Wheel radiuses of 0,6 m, the height of escarp of 0,9 m are worked out, special attention being paid to design data of the calculations schemes (in accordance with the scale). The geometrical parameters of the machine body and distances between wheel axles are also similar to the real construction (2,5 is a wheel radius between two forward axles and two rear axles; 3,5 is wheel radius between second and third axles).

Vertical reactions on the single wheels of such vehicles usually are similar, that is $R_1 = R_2 = R_3 = = R_4 = 0{,}25\ G_a$. The distinctions do not usually exceed 5-10 %. It is necessary to consider the effect on the static weight distribution of the location of the front power unit which is more demanded now, and the absence of the troops in the compartment at the moment of overcoming obstacles, the total weight of them reaches 1000 kg (10 people equipped with military equipment, 100 kg each), the final distribution of the static vertical loads on the axles (curb weight of the machine is 12 tons) is taken:

$$R_{1c} = 0{,}27\ G_a\ ; \quad R_{2c} = 0{,}26\ G_a\ ; \quad R_{3c} = 0{,}24\ G_a\ ; \quad R_{4c} = 0{,}23\ G_a\ .$$

The first stage of the machine movement with small velocity through the escarp, the height of which is 0,9 m, is starting on the contact of the top edge of the bow part of the body with the top of the escarp. The armoured bow, which usually has corresponding inclination of the lower front part for rising protective properties under action of draft forces F_{T1}, F_{T2}, F_{T3} and F_{T4} provides sliding of the front part of the body on top of the escarp while touching the vertical wall of the escarp by the front wheels (figure 1). The total force $F_{\Sigma 1}$ of pressing the first axle wheels to the wall at that moment is equivalent to

$$F_{\Sigma 1} = F_{T1} + F_{T2} + F_{T3} + F_{T4} = (R_1 + R_2 + R_3 + R_4)\ \varphi_x = \varphi_x\ G_a\ , \qquad (1)$$

where R_1, R_2, R_3, R_4 – vertical reactions of the support surface on wheels of the corresponding axle of the machine with consideration of their redistribution from reactive moments on the body from driving axles (wheels); φ_x – coefficient of adhesion of the wheels with the support surface.

Reactive momentum from forces $F_{т1}$, $F_{т2}$, $F_{т3}$ и $F_{т4}$ forms unload action for wheels of two front axles, and the sum of reactive momentum T_p from that forces in the first approach is equal to

$$T_p = (F_{т1} + F_{т2} + F_{т3} + F_{т4})\, r_д = \varphi_x(R_{1c} + R_{2c} + R_{3c} + R_{4c})\, r_д = \varphi_x\, G_a r_д\ . \quad (2)$$

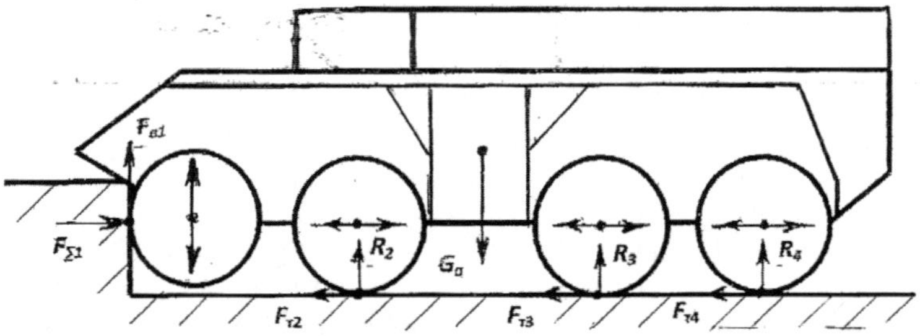

Figure 1. Scheme of the forces on machine with wheel formula 8x8 at the moment of starting to overcome the escarp (the first driving axle wheels start an upward movement; pair of the forces for calculation the reactive momentum are applied on axles)

It is possible to consider with sufficient accuracy, that it gives the reduction of vertical reactions on wheels of the front axle on ΔR_1, on wheels of the second axle – on $0{,}5\Delta R_1$. Accordingly, on wheels of the rear axle it gives the increase of vertical reactions on ΔR_1, on wheels of the third axle – increase is $0{,}5\Delta R_1$. That is why the size ΔR_1 may be determined for this calculation scheme from such dependences:

$$T_p = 8{,}5\Delta R_1\, r_д + 3{,}5\Delta R_2\, r_д = 8{,}5\Delta R_1\, r_д + 1{,}75\Delta R_1\, r_д \approx 10\Delta R_1\, r_д$$

$$T_p = \varphi_x\, G_a\, r_д = 10\Delta R_1\, r_д;\quad \Delta R_1 = 0{,}1\, \varphi_x\, G_a\ . \quad (3)$$

In case of asphalt surface $\varphi_x = 0{,}8$, we get $\Delta R_1 = 0{,}08\, G_a$. Accordingly, the dynamic distribution of vertical load on machine axles during of overcoming of escarp will be the following:

$$R_1 = 0{,}19\, G_a;\quad R_2 = 0{,}22\, G_a;\quad R_3 = 0{,}28\, G_a;\quad R_4 = 0{,}31\, G_a.$$

With the total force of pressing to the front wheels to the wall of escarp $F_{\Sigma 1} = \varphi_x\, G_a$ (2), on the front wheels it $F_{в1} = \varphi_x^2 G_a = 0{,}64\, G_a$ is guaranteed, vertical force that is enough for beginning its moving upwards ($F_{в1} = 0{,}64\, G_a > R_1 = 0{,}19\, G_a$). After the front wheels come onto the horizontal surface, the total force $F_{\Sigma 1}$ will be reduced because the forward pushing force from the wheels of the front axle disappears (figure 1). But the pushing force from the driving wheels of the second, the third and the forth axles is enough for reliable pressing the wheels of the front axle to the vertical wall of the escarp and its continuation for confident moving upwards:

$$F_{в1} = (0{,}22\, G_a + 0{,}28\, G_a + 0{,}31\, G_a)\, \varphi_x^2 = 0{,}64 \bullet 0{,}81 G_a = 0{,}52\, G_a > 0{,}19\, G_a = R_1\ .$$

Then hanging out of the second axle wheels takes place (figure 2), which leads to the redistribution of the vertical load $0{,}22\, G_a$ on the stay for supporting the three other axles. This redistribution according to the geometry axles arrangement on the machine length may be taken as following:

$$\Delta R_1 = 0{,}12\, G_a;\quad \Delta R_3 = 0{,}06\, G_a;\quad \Delta R_4 = 0{,}04\, G_a\ .$$

Accordingly, the real vertical loads on the wheels of the machine at that moment are the following:

$$R_1 = 0{,}31\, G_a;\quad R_2 = 0;\quad R_3 = 0{,}34\, G_a;\quad R_4 = 0{,}35\, G_a.$$

Moving of the front driving wheels upwards along the vertical wall of the escarp and then forward will continue during such hanging out because the necessary condition should be fulfilled:

IOP Conf. Series: Materials Science and Engineering **386** (2018) 012020 doi:10.1088/1757-899X/386/1/012020

$$F_{B1} = (0,34\ G_a + 0,35\ G_a)\varphi_x^2 = 0,64 \bullet 0,69\ G_a = 0,442\ G_a > 0,31\ G_a = R_1\ .$$

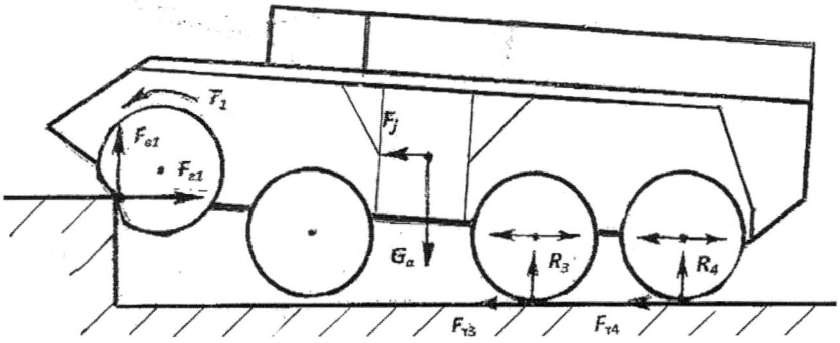

Figure 2. Scheme of the forces on the machine with wheel formula 8x8 at the final moment of moving of the front axle wheels up to the vertical wall of the escarp

At the final moment of moving the front axle driving wheels up the vertical wall of the escarp (figure 2) the front wheels by the grousers engage on the rib of the escarp, thus leading to 20-30% increase of its adhesion coefficient [2, 3]. Momentum T_1 , which acts on these wheels, and forces inertia F_j (because of some decrease of the speed moving of the machine) in combination with forces F_{T3} and F_{T4} leads to confident moving of the machine forward.

The next stage of overcoming a big escarp by machine 8x8 is moving along its vertical wall of the second axle driving wheels. To this moment the front axle driving wheels are already situated on the upper horizontal ground of the escarp and create draft force F_{T1}. The wheels of the second axle hung out at maximum, the vertical reaction on them is absent (figure 3).

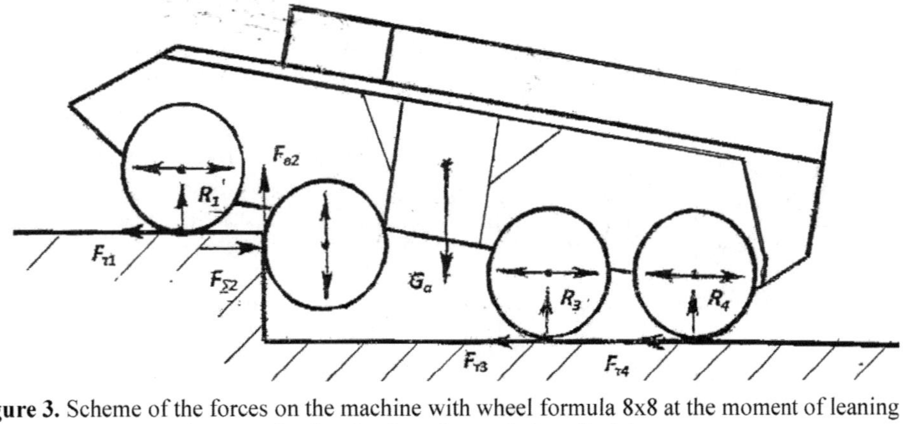

Figure 3. Scheme of the forces on the machine with wheel formula 8x8 at the moment of leaning of the second axle wheels to the vertical wall of the escarp

In these conditions the vertical movement of the second axle wheels on the escarp wall is performed without any difficulties because even on their maximum upwards moving the load on them does not exceed 0,25-0,4 G_a and force $F_{\Sigma 2}$ of pressing to the wall by these wheels is

$$F_{\Sigma 2} = F_{T1} + F_{T3} + F_{T4} = \varphi_x G_a = 0,8\ G_a\ .$$

Accordingly, force F_{B2}, which moves them up, at maximum is able to reach of the size $F_{B2} = \varphi\ F_{\Sigma 2} = 0,64\ G_a > 0,4\ G_a$, that is more than enough.

The second axle wheels moving upwards are over with the situation, which is similar to fig. 2 for the front axle, when the wheels rest on the escape edge. In that position it is possible to hang the front axle wheels because the bow part of the machine is raised upwards strongly. That means approaching the vertical load of the second axle wheels to 0,4-0,5 G_a level. At the final moment of moving the

IOP Conf. Series: Materials Science and Engineering **386** (2018) 012020 doi:10.1088/1757-899X/386/1/012020

second axle wheels upwards on the wall of the escarp the grousers of the tires catch on the rib of escarp, what leads to increasing their adhesion coefficient. Momentum T_2 , which acts on these wheels, and force of inertia F_J (because of some decrease of the speed moving of the machine) in combination with pushing forces F_{T3} and F_{T4} assist to confident moving of the machine forward.

With the third axle wheels pressing to the vertical wall of the escarp it starts the third stage of the process, which scheme is introduced in figure 4. Very likely, the third axle wheels will be hanging also how it was during overcoming the escarp with the second axle. At the initial stage of moving the third axle wheels upwards it is no of any difficulties. Practical absence of the essential vertical load on them leads to its moving upwards on the wall easily because of pressing force FΣ3 from drag forces FT1, FT2 and FT4 (as it was illustrated in the example of the second axle) is more than enough, so force $F_{B3} >> R_3$.

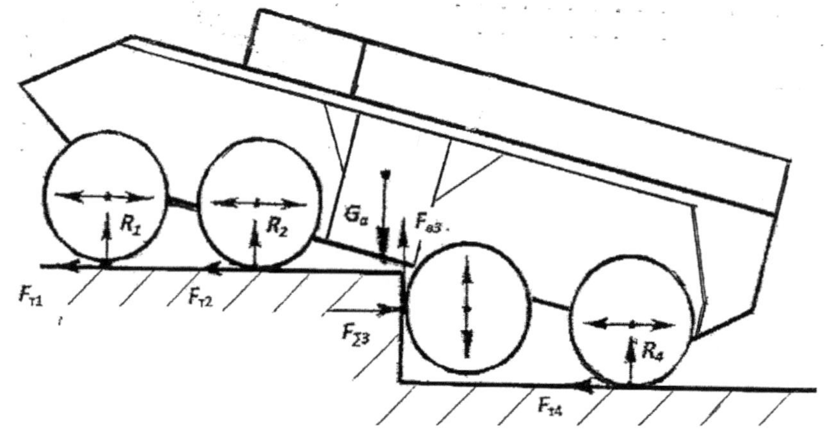

Figure 4. Scheme of the forces which act on the machine with wheel formula 8x8 at the moment of leaning the third axle wheels against the vertical wall of the escarp

Problems arise when lifting the third axle wheels to the stop in the vertical suspension travel limiter due to the beginning of hanging the rear axle wheels (figure 5). The vertical load on the wheels of the third axle is thus approaching its possible maximum: 0.5 Ga. In this position crucial influence to overcome obstacles exerts a torque on the wheels of the third axis, which is due to the high grip of these wheels with the edge of the escarpment provide a significant increase in push force to the pull forceF_{T1} and F_{T2} from the front axles. Some additive also gives an inertial effect of F_j which will be the result of small decrease in the speed of horizontal movement along the obstacle at this moment.

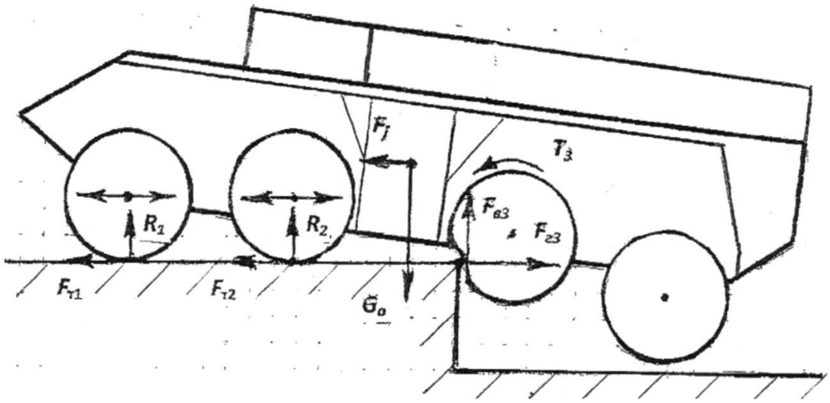

Figure 5. The scheme of forces acting on the machine with a wheel formula 8x8 when lifting the wheels of the third axis until it stops in the vertical suspension travel limiter (one of the most difficult modes of movement on the escarpment)

After the entry of the third axis wheels on the upper platform of the escarp, the entry of the rear (fourth) axis wheels on it is not particularly difficult. It is because in this situation machine, the amount of tractive effort F_{T1}, F_{T2} and F_{T3} on wheels three front axles located at the top of the escarpment, significant, and the drive wheels of the fourth axle, which is in the posted state when substantially horizontal position of the machine frame, to enter onto a ledge, the height of which will actually be significantly smaller than the radius of the wheel. At the same time, at the initial moment of entry to the escarp edge, the vertical loads on the rear axle wheels are minimal, since they are posted.

The obtained theoretical results are confirmed by the relevant experiments of foreign researchers. It should be noted that the presence of the foreign all-wheel drive eight-Wheeler hydropneumatic suspension allowed in the process of moving along the obstacle to reduce the longitudinal rolls of the machine, but this practically did not affect the course of the process in terms of facilitating the overcoming of the escarp, the height of which, as follows from official publications, during the tests was 0,9 m (figure 6 – figure 9), which corresponds to the taken basis for the analysis of the calculation scheme of the situation.

Figure 6. Machine *General Dynamic* overcomes escarp 0,9 m by the front wheels

Figure 7. Second axle wheels entry on upper ground of the escarp, front wheels hang without contact with the supporting surface

Figure 8. The third axle wheels storm the rib of the escarp, the vertical load on them as may be seen from the axle picture is at maximum

Figure 9. Three front driving axles entry on the upper ground of the escarp, the rear axle wheels hang in the air

The theoretical analysis and the experimental data obtained from the literature allow us to draw the following conclusions:

1. It is confirmed that an all-wheel drive machine 8x8 have improved capabilities to overcome the biggest obstacles of rectangular type of the escarp compared to machines with wheel formula 4x4. To overcome escarpment of height of 1.5-1.7 radius of the wheel by such machines, which is almost twice as high for cars with a wheel formula 4x4 is real.

2. Geometrical parameters of the bow part of an all-wheel drive machine 8x8 can have a significant impact on the success of overcoming obstacles exceeding the radius of the wheel. From this position, the placement of protruding elements such as towing devices or lighting devices on the lower panel of the bow is undesirable, as they may prevent (see figure 1) sliding of the front of the machine on the top edge of the escarpment.

IOP Conf. Series: Materials Science and Engineering **386** (2018) 012020 doi:10.1088/1757-899X/386/1/012020

References

[1] Peskov V I, Serdiuk V I and Serdiuk A E 2009 *Improving of exploitation qualities of automobile* (Nizhny Novgorod: Nizhny Novgorod State Technical University n.a. R.E. Alekseev) p 135
[2] Ageikin J S, Volskaya N S and Chichekin I V 2000 *Passability of automobile* (Moscow: Moscow State Industrial University) p 141
[3] Ageikin J S and Volskaya N S 2008 *Theory of automobile* (Moscow: Moscow State Industrial University) p 318

IOP Conf. Series: Materials Science and Engineering **386** (2018) 012021 doi:10.1088/1757-899X/386/1/012021

Problems of the application of intelligent driver assistance systems on a single-track vehicles

V V Gaevskiy, A M Ivanov

Moscow Automobile and Road Construction Technical State University (MADI), Moscow

vit-life@rambler.ru, ivanov-am@madi.ru

Abstract. This article considers some problems and features of the application of modern systems of driver assistance on single-track vehicles ("STV"). It is also mentioned the classification of intelligent driver assistance systems and design features of the "STV" that require an improvement for their successful application on such vehicle. There is also described perspective of systems' development that helps the "STV" driver. Also perspective the "STV" types with high safety construction are considered in this article.

Introduction

In January 2018 there were discussed essential improvement questions regarding the safety of motor transport in Europe during the regular conference of the European Association of motorcycle manufacturers (ACEM - European Association of Motorcycle Manufacturers). There were discussed problems regarding the environmental safety improvements, active and passive safety of motor vehicles, the development of intelligent driver helping systems ("IDHS") that were adapted for their installation on a single-track vehicles ("STV"). Also there were considered the interaction of the "STV" with other participants on public roads, the development and improvement of training systems for drivers of motor vehicles and other questions related to the safety of the "STV".

One of the conclusions made in ACEM conference resolution is a need of the intensive development of the "IDHS" for the "STV". This measure will significantly reduce risks of road accidents that occur with motor vehicles, and in case of getting the "STV" in an accident - it may reduce the severity of the consequences.

Dynamics of road accidents and number of the "STV" on roads of the EU and Russia.

According to the latest data of IRTAD (International Road Traffic and Accident Database - an international database of road traffic and accidents), accidents of two-wheeled vehicles decreased by 44 % from 2000 to 2014 years. In the period from 2010 to 2014 years, the number of deaths among motorcyclists decreased from 4304 to 3561 people (that is 17.3% less). In the same period the number of deaths among scooter drivers decreased from 975 to 622 people (that is 36.2% less). At the same time the number of the "STV" in the EU increased by 5.9% from 2000 to 2014 time period and amounted to 36.1 million by the end of 2014 year.

Research studies and works show us that the reduction of accidents and the severity of the consequences with the participation of the "STV" due to the increased number of the "STV" equipped with driver assistance systems (anti - lock systems, etc.) is the beginning of its mass application during the period of 2008-2010 years.

According to Road Police data in Russia from 2012 to 2015 years number of deaths among all "L" vehicle categories decreased from 2033 to 1534 people (that is 24.6% less). But the number of the "STV" decreases each year at about 0.5-1 % in Russia. By the end of 2015 year there were officially registered 2.25 million L3 - L5, L7 categories of motor vehicles.

If we check figures of the reduction of death occasions in the EU and in Russia, it may seem that the situation in Russia even better than in Europe. Unfortunately it is not so. To be correct comparing the safety level of the "STV" in Russia and in the EU, it is important to consider not absolute figures, but relative ones. Let's count the number of deaths per 10,000 the "STV" units. In the EU it is 1,2 human per 10.000 the "STV" and in Russia - the number is 6,8 humans. Thus, driving motorcycle vehicles in Russia is 5,6 times more dangerous than in European Union.

The situation in Russia is even worse if we compare the number motorcyclists died during road accidents with the number of automobile drivers died in the same conditions.

The specific number of deaths per one registered motorcycle is twice more than the specific number of automobile drivers died during road accidents (3,2 per 10 thousand cars of the M1 vehicle's category).

If we compare death quantity of motorcyclists per one kilometer with the quantity of deaths of automobiles' drivers and passengers we'll find out that motorcyclists die 10 times more often!!!

The "STV" is the most dangerous type of vehicles in the world and especially in Russia!

Design features of single-track vehicles.

Advanced driver assistance systems (ADAS): adaptive cruise control (ACC), autonomous emergency brake systems (AEBS) and others, that were primarily designed for automobiles, cannot be installed on a motorcycle without necessary additional adaptation actions to its design features. Intelligent driver assistance systems require special methods and specific technical solutions to optimize their capacity to operate on the "STV" to improve road safety [1,2].

It is also very important to point out that these systems will require the development of appropriate human-machine interfaces (HMI). The "HMI" should minimize driver's distraction and should be specifically designed for the "STV". For example, messages should be prioritized to make general notices be replaced by security warning notifications.

The main design features of the "STV":

- insufficient stability during driving mode with low speed and unstable equilibrium while standing still; (Figure 1)
- due to the driver's and passenger's open sitting position on the "STV" they are directly exposed to negative external influences: snow, rain, wind, dust, etc.
- "STV" open construction requires the mandatory use of protective devices according to the regulations of all countries: a helmet, at least; "shells"; overalls; crags, etc.;
- motorcycle fit: driver's offset and inclination affect the vehicle's steerability, stability and fatigability of the driver during moving mode; (Figure 2)
- "STV" motion feature during its turn actions - with a large inclination, with the displacement position of the contact point; (Figure 3) [3].
- "STV" specific control units and control systems;
- "STV" are vehicles without bodies, therefore they have the worst passive safety during an accident.

For further consideration of perspectives and problems of the "IDHS" application for the "STV" we'll describe their simple classification.

IOP Conf. Series: Materials Science and Engineering **386** (2018) 012021 doi:10.1088/1757-899X/386/1/012021

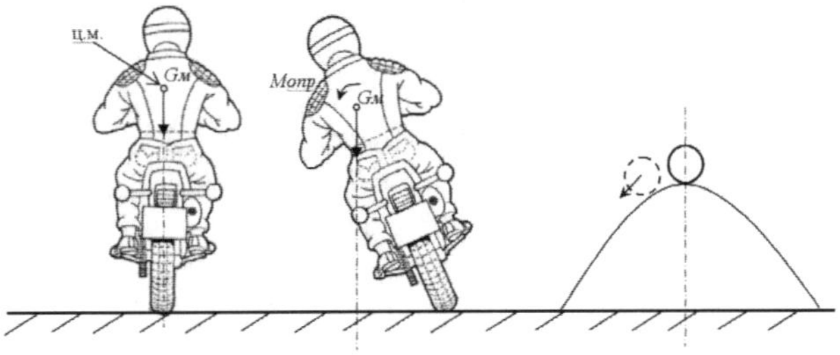

Figure 1. Unstable equilibrium position of the "STV" while standing still and driving with low speed.

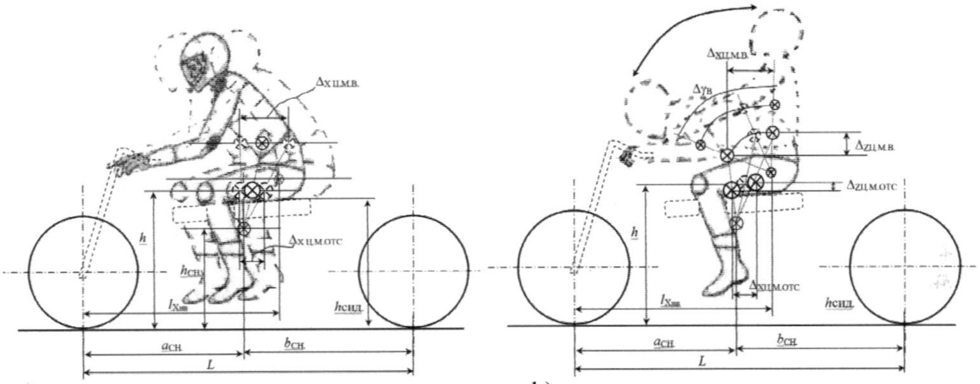

a) b)

Figure 2. Types of driver's movements on the "STV" and displacement of general gravity center at the same time. a-longitudinal; b - inclination.

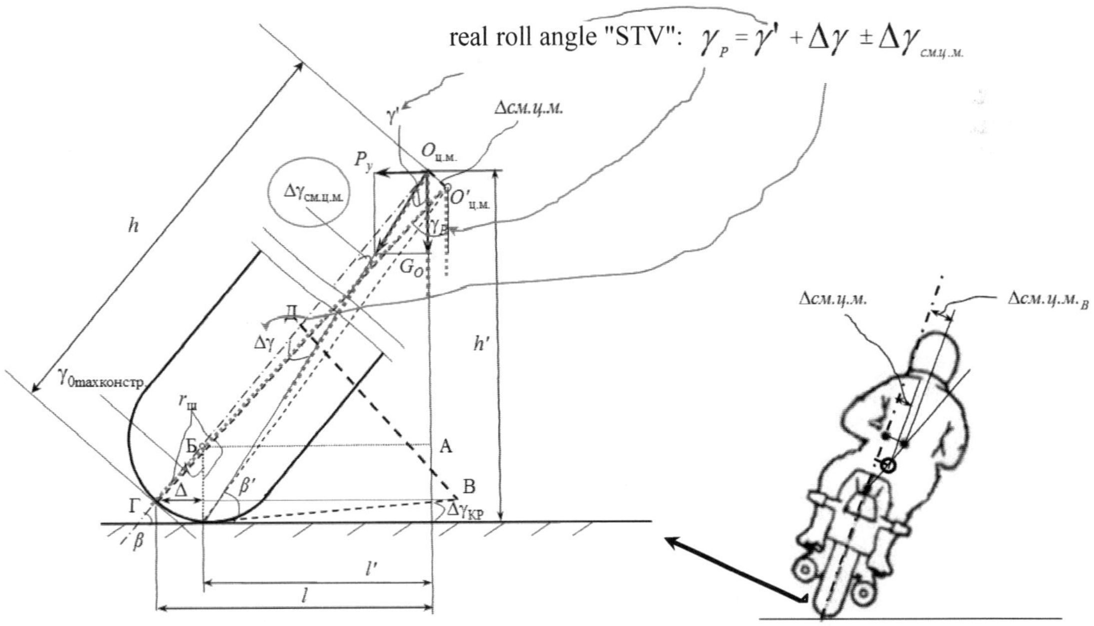

Figure 3. Motion during a turn mode using wide tires (taking into the consideration driver's displaced position).

Influence rate that affects the motion process of the vehicle:

- information, for example: marking recognition, navigator, information about "dead zones", adaptive lighting systems, etc. [4,5,6]

These systems only inform the driver without interfering in the motion process of the vehicle.

- executive and information, for example: AEBS;

At first, such systems warn the driver about the occurrence of a dangerous situation and necessity to take all important measures. If the driver (for any reason) does not react to warning signals, the system itself interferes into the moving process of the vehicle (in this case it slows down).

- automatically actuated, such as: ABS, ESP, rain sensor, etc.

These systems always work in automatic mode without the driver's participation. But they begin to influence actively vehicle's driving mode only after the occurrence of a dangerous ("ABS") or complicated (rain sensor) road situation.

- Autonomous and robotic, switched on by the driver itself (partial or complete replacement of the driver), for example: adaptive cruise control, lane keeping system, robot.

The purpose of these systems is partly (adaptive cruise control) or fully (robot driver) operation of driver's functions during vehicle's driving mode.

By purpose (scope):

- during braking mode: ABS, DSC, ASC, HDC, EBD, CBC, EBA, BAS, PBA, EPB, AEBS;

Systems of this group help the vehicle to brake safely during different road situations.

- during turn mode: ESP, DAS, AFS;

This "IDHS" group affects the vehicle's steering system to keep safe driving direction.

- during acceleration mode: ETS, ASR, EDS;

They provide a reliable and safe acceleration of the vehicle.

- universal: LKSS, SA, ACC, PDS, IPAS, etc.;

These "IDHS" affect several control units and can influence all vehicle's driving modes: acceleration, braking, changes of direction etc.

- during parking mode: PTS, AVM, TPMS;

Specialized systems that assist the driver during vehicle's parking mode in narrow space, completely or partially replace the driver's functions.

- comfort improvement: rain sensor, climate control, etc.;

These systems improve the comfort of vehicle control. They also improve traffic safety. As research studies have shown that the accumulation of driver's fatigue increases the probability of getting into an accident.

- after an accident: eCall (emergency warning system), seat belts, active head restraints, airbags.

Very important helping systems. These systems help to increase chances of human survival after an accident.

Due to the design features of the "STV" practically none of the above mentioned systems may be applied without its prior adaptation, and some systems can't still be used for the "STV" at all.

Intelligent driver assistance systems for automobiles that are not used on the "STV".
Some systems can't be applied for the "STV" and the reason is that there is simply no place for the installation. Some systems are not used because there is no need. And some systems can't yet be used on the "STV" due to control units action problems. For example: what happens to a motorcycle if the system that prevents the hitting a pedestrian suddenly jerks the steering wheel to avoid a crash in case of stop failure? Possibly the vehicle will fall on the side and during the sudden braking it will simply overturn.

The list of systems that are not still used during serial production of the "STV":

IOP Conf. Series: Materials Science and Engineering **386** (2018) 012021 doi:10.1088/1757-899X/386/1/012021

- alert system "dead zones";
- parking assistance systems (PTS, AVM, TPMS);
- Autonomous and robotic (partial or complete driver replacement system) - adaptive cruise control (ACC), lane keeping system (LCSS), robotic driver, etc.;
- help system after road accident: eCall (emergency alert system), safety belts, active head restraints, airbag;
- active steering systems: (DAS, AFS);
- body internal comfort systems (rain sensor, climate control, etc.).

Intelligent driver assistance systems installed during serial production of the "STV".
In present time the most popular "IDHS" that is applied for the "STV" is the "ABS" (Anti-lock braking system). The matter is that blocking of wheels during brake mode is the most dangerous action for single-track vehicle. The stable condition of the "STV" during high speed moving mode is provided by gyroscopic stabilizing moments of its wheels. If they are blocked (upon instant termination of the stabilizing moments) the stability of the "STV" becomes similar to its static mode (see Figure 1). As a consequence the "STV" instantly falls down. However, due to the "STV" design features it was not possible to apply the "ABS" from the automobile. It had to be improved significantly (Figure 4). The motorcycle has a separate brake drive on the front and rear wheels and two brake control units. Therefore, it has not one master brake cylinder, but two.

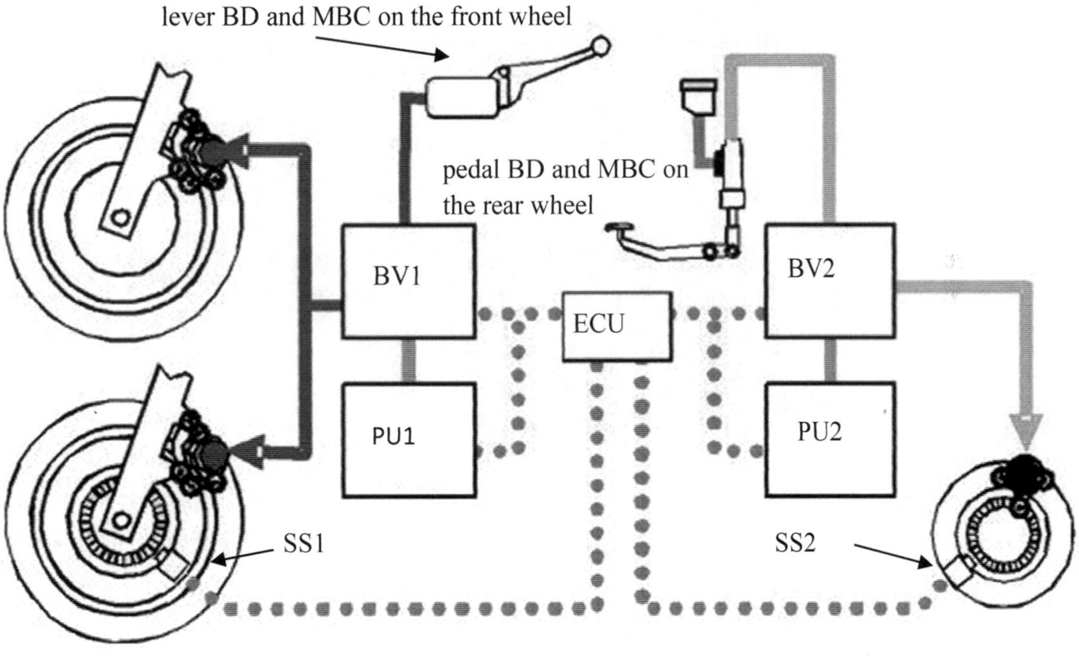

Figure 4. ABS schematic drawing for "STV".

BD - brake drive, MBC-the muster brake cylinder, ECU- ABS electronic control unit, BV-block of valve, PU-power unit, SS-speed sensor, 1 - front wheel operating scheme, 2 - rear wheel operating scheme.

Separate brake drive on the front and rear wheels and two control units reduce efficiency of braking process of the "STV". Research works have shown that using a single brake control unit improves accuracy and efficiency of the braking system. At present time many manufacturers began to install special systems on motorcycles that help the driver during braking mode. These systems include: CBS (Combined Brake Systems) a combined brake system and UBS (Unified Brake System) unified brake system. The feature of these systems is partial (CBS) or complete (UBS) substitution function of second brake control unit. When the driver operates only one brake control unit (e.g. a lever), the system starts to brake the second wheel itself. The main problem of such systems is that during the "STV" braking mode weight per front and rear wheels is redistributed in a complicated way (Figure 5).

Additional feature of the "STV" braking system is high probability of a complete distribution of weight per front wheel (possibility of rolling forward during braking mode). If the "STV" has lighter weight than the possibility of such braking process is much more probable to happen.

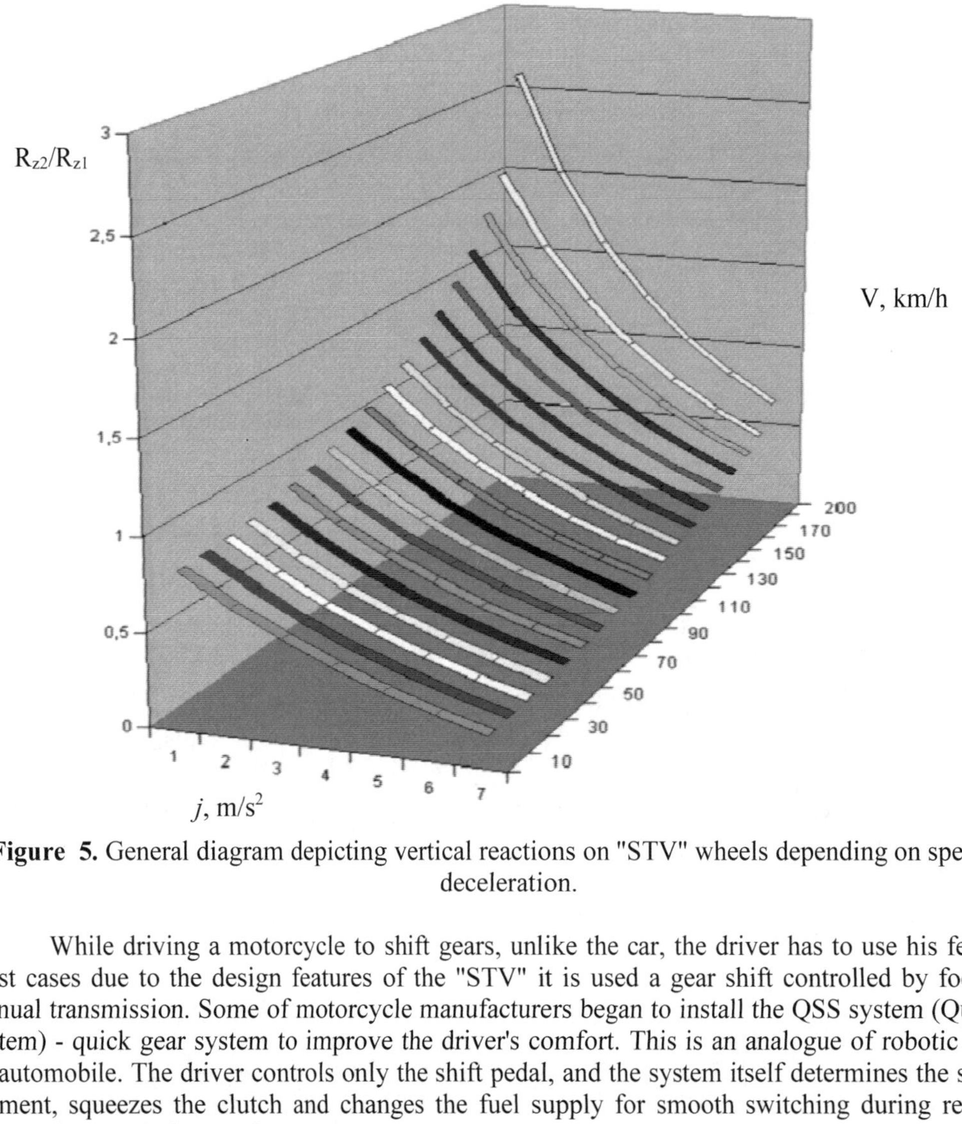

Figure 5. General diagram depicting vertical reactions on "STV" wheels depending on speed and deceleration.

While driving a motorcycle to shift gears, unlike the car, the driver has to use his feet. As in most cases due to the design features of the "STV" it is used a gear shift controlled by foot with a manual transmission. Some of motorcycle manufacturers began to install the QSS system (Quick shift system) - quick gear system to improve the driver's comfort. This is an analogue of robotic gearshift on automobile. The driver controls only the shift pedal, and the system itself determines the switching moment, squeezes the clutch and changes the fuel supply for smooth switching during reduced or increased transmission mode.

Another critical and very dangerous "STV" driving mode is sliding during the turn mode. Unlike the automobile for which the beginning of turnover during a turn mode is more dangerous, for the "STV" the most dangerous is the beginning of a sliding moment. The matter is that during a turn mode the "STV" moves steadily only when the projection of the resulting gravity and centrifugal forces passes through the contact point of the wheel with the road (see Figure 3). If the lateral slip situation begins during the "STV" turn mode real turning radius increases suddenly, as a result centrifugal force reduces and the "STV" falls dawn. To prevent this situation, SCS (Slide control system) is used a sliding control system during the turn mode. It fixes the beginning of the slip and partially perform braking with the help of sensors, besides it reduces the fuel supply to engine cylinders [7].

Increasingly, the "IDHS" systems are used for modern motorcycles. They control the acceleration process and particularly prevent the occurrence of the rollover situation - front wheel shifting during acceleration. Unlike an ordinary automobile, the "STV" may easily roll over back with a shift acceleration, which is very dangerous. The feature of such systems (LIF (Lift control system) - front wheel lift control system; LCS (Launch control system) - start control system; TCS (Traction control systems) - traction control system) includes the determination of the beginning of the rollover and torque reduction on the drive wheel, to prevent this process. In addition, the TCS system determines the start of the motorcycle's driving wheel slip, reduces torque and ensures its acceleration with maximum efficiency and safety.

Another problem that may happen during driving a motorcycle at speeds from 40 to 80 km/h is the self-oscillation of the front controlled fork, so-called "wobbling" of the front part of a motorcycle. The EVD (Electronic vibration damper) system - electronic vibration damper handlebar is designed to eliminate such oscillation. Many of the motorcycles have much more simple system as a common damping (transverse damper) on the steering fork.

Recently BMW Company has announced the readiness to install the eCall system as an additional option on their motorcycles.

Perspective types of the "STV" and the intelligent driver assistance systems that are used for them.

Currently electric bicycles and small electric scooters are very popular among the customers as they are environmentally friendly transport. The advent of electrically driven "STV" gave a possibility to apply the "IDHS" to compact and light vehicles that previously could not be fitted: such as the ABS and the CBS.

In Europe the quantity of stable three wheeled "STV" increases. They are safer and more comfortable during the operation, but at the same time they have the advantages of a single-track transport: compact dimensions and good steering parameters (Figure 6).It is also possible to increase the stability of the "STV" at low speeds and during a stable mode, using retractable landing gear [8].

Figure 6. Example of sustainable and compact three-wheeled scooter.

However, we believe that the most promising and safe "STV" should be a new type of compact vehicle - Body single-track vehicle ("BSTV") (Figure 7). The presence of the body relates it into a class of safer vehicle, comparable to the automobile [9, 10]. Passive safety of the "BSTV" becomes similar to the automobile. Besides it is possible to install the internal systems of driver assistance, such as seat belts, airbags, navigation systems, etc. Protection from the external negative factors increases the comfort and safety of operation and makes it possible to operate the "BSTV" all year, regardless of weather conditions.

On the "BSTV" it may be installed absolutely all the same "IDHS" systems, as on a conventional automobile. Besides the "BSTV" is more compact, economical and environmentally friendly compared to a conventional automobile, and therefore more perspective as a city vehicle.

Figure 7. Examples of "BSTV".

Summary.

Single-track vehicles, being the most dangerous in the world are least equipped with intelligent driver assistance systems. Due to the design features and traffic peculiarities, in order to install the "IDHS" systems on the "STV", the substantial adaptation actions of each system are required. Besides, some systems cannot be installed on single-track transport at all. On compact types of the "STV", such as bicycle and scooter, no driver assistance systems are installed at all.

Further improvement of the "STV" traffic safety is possible only with the introduction of the new types of the safe "STV" and new driver assistance systems, as well as the expansion of the application of existing "IDHS" on a larger number of the "STV".

References

[1] Ivanov A M Basics of construction of modern car: a schoolbook for universities. / A M Ivanov, A N Solntcev, V V Gaevskiy; P N Klukin, V I Osipov, A I Popov / — M.: Engineering Research and Education Center «SMART», 2017. — 348 p.: ill.
[2] Ivanov A M Cars: The theory of operational properties: a schoolbook for student institutions of

higher professional education. / A M Ivanov, A N Narbut, A S Parshin, A N Soltcev, V V Gaevskiy; edited by A M Ivanova. // M.: Publishing center "Academy ", (second edition) - 2014. - 176 p.
[3] Gaevskiy V V Stability of single-track vehicle movement on wide tires in turn. Auto Transport Enterprise. 2015. № 4, p. 26-28.
[4] Sergey Sergeevich Shadrin, Andrey Mikhailovich Ivanov. Algorithm of Autonomous Vehicle Steering System Control Law Estimation while the Desired Trajectory Driving // ARPN Journal of Engineering and Applied Sciences. 2016. Vol. 11, No. 15. P. 9312-9316
[5] Shadrin S S, Ivanov A M, Nevzorov D V Autonomous wheeled vehicle as part of intelligent transport systems // Natural and technical sciences. 2015. release 6(84). p. 309-311.
[6] Ivanov A M Development of a system of interobject interaction of intelligent vehicles / A M Ivanov, S S Shadrin // Izvestia VolgGTU. Series «Ground transportation systems». release 7: interuniversity. sat. sci. art. / VolgGTU. – Volgograd, 2013. – № 21 (124). – p. 74-77.
[7] Gaevskiy V V Perspective ways to improve the safety of single-track vehicles. / V V Gaevskiy, M S Podolskiy // Journal of Automotive Engineers. - 2015. № 1 (90). - p. 36-39.
[8] Gaevskiy V V Retractable chassis for motorcycle. / V V Gaevskiy, E.R. Sultanova // Auto Transport Enterprise. - 2016, № 11. - p. 37-39
[9] Gaevskiy V V Single-track vehicle with constantly-vertical wheels. Patent for utility model № 133498. Registered in the State Register of Utility Models RF from 20.10.2013 g.
[10] Gaevskiy V V Cover one-track (COT) - perspective view of the city compact single-track vehicle / V V Gaevskiy // Proceedings of Nizhny Novgorod State Technical University. R E Alekseeva.- NGTU them. R E Alekseeva. - Nizhny Novgorod, 2014. № 4 (106). materials of the 87th International Scientific and Technical Conference "Operational Safety of Motor Vehicles". - p. 219-223.

IOP Conf. Series: Materials Science and Engineering **386** (2018) 012022 doi:10.1088/1757-899X/386/1/012022

Rated definition of daily fuel consumption of a city shuttle bus

V I Peskov, N A Kuzmin, A N Kuzmin

Nizhny Novgorod State Technical University n.a. R.E. Alekseev, 24 Minin Street, Nizhny Novgorod, Russia

Abstract. Rated definition of fuel consumption by vehicles becomes the new direction of improvement of transportation processes. Calculations of daily fuel consumption by bus PAZ with the diesel engine used on city routes of Nizhny Novgorod with use of specifications of a definition technique of specific fuel consumption depending on extent application of engine power are executed. Results are compared with the existing normative indicators.

The specifications [1] made concerning a method of calculation of fuel consumption of the moving wheel car allow to calculate such useful properties of vehicles as daily fuel consumption on a specific route of the movement and also to make impact assessment on some factors, for example, for the purpose of searching ways of efficiency increase by using city buses.

For the theoretical analysis we will consider a real situation with shuttle buses of middle class buses PAZ-3204 with Cummins diesel engines which are widely used in Nizhny Novgorod. Technical characteristics of these cars are presented in table 1 [2]. The average daily run of such moving buses in city conditions makes 240 km, i.e. if the one way route length equals 20 km, the bus makes 12 trips per day from one final stop to another.

Table 1. Technical characteristics of the bus PAZ-3204 equipped the Cummins engine

Parameter	Value
Mass of the equipped bus, kg	5055
Width-height, m	2,41 x 2,88
Maximum engine power (kW)	104
Angular speed of the crankshaft (rad / sec)	272
Maximum engine torque (N*m)	502
Angular speed of the crankshaft (rad / sec)	136-178
Gearbox ratios:	4,65; 2,6; 1,53; 1,0; 0,77.
Main gear ratio:	4,55
Streamlining coefficient	0,75
Tyres	245/70*R*19,5
Control fuel consumption, l / 100 km (at a speed of 60 km/h)	16,3

For calculations of traveling fuel consumption of the moving bus with use of the known formulas of the theory of the cars [1, 5] it is necessary to know specific fuel consumption at maximum power g_{ep} of Cummins engines. In this case quite correct result can be received on the basis of trial operation of completely loaded cars "Gazelle Next" equipped with Cummins diesel engines. It is known that with traveling speed of 60 km/h on 4th (direct) gear control fuel consumption was 8,5 l/100 km, at speed of 80 km/h it was already 10,3 l/100 km. These data in combination with technical characteristics of the specified cars allow to determine g_{ep} in accordance with the mentioned formulas, and the obtained rated result (g_{ep} = 227,5 ± 0,5 g/kW.h) is almost identical to both options of the movement which affirms its reliability. In favor of such conclusion also calculation of g_{ep} proceeding of control fuel consumption (see table 1) the bus PAZ-3204 testifies.

One of the routes analyzed in this article has 16 stops. Therefore, for one trip along the route a driver carries out accelerations from zero speed to admissible 60 km/h in city conditions with gear shift use at least 15 times. At the same time, as observations showed, drivers use the first gear for start-off from the place very seldom. Rather high torsion torque of the diesel allows to start from the second gear. It is confirmed also by dynamic characteristics of the car which testify that the maximum ability rating on the second gear (a gear ratio 2,6), equals 0,18-0,19 when loading the bus by passengers up to 60-75%, it allows to make dispersal from the place with acceleration up to 1,5-1,7 ms^{-2}. For 12 daily trips around the specified route of such start-off from the place at least 180 movements with acceleration on the third and fourth gears will be made.

The mass of the moving bus constantly changes because of passengers getting in and out at stops. In calculations it is necessary to operate with some average value. During its movement on a route average statistical data of changing passengers number inside the bus (figure 1) on this specific route allow to define average bus loading. From the obtained data it turns out that the average number of passengers in n_m traveling by this bus route at approximately identical run between stops can be accepted as equal 19. Therefore the average full mass m_{fm} of the bus with passengers and the driver is equal

$$m_{fm} = m_o + m_p\,(n_m+1) = 5055 \text{ kg} + 68\cdot 20 \text{ kg} = 6415 \text{ kg},$$

where m_o –mass of the equipped bus, kg; m_p –mass of a passenger (68 kg [3]).

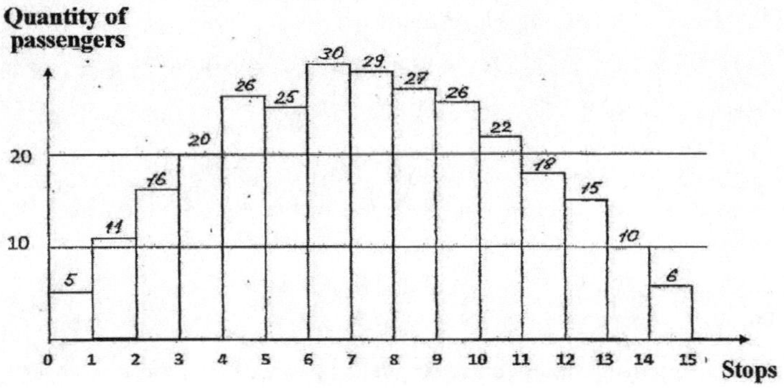

Figure 1. Change of the passenger number inside of the bus in the process of the movement on a route from an initial stop to the final one.

The first stage of acceleration of the bus from the place begins with preliminary promotion of an engine flywheel to angular speed 170-190 s^{-1} and the controlled driver of a clutch slip by the included second gear in a gearbox. The stage lasts 1,0-1,5 s and continues due to inertia of a flywheel acceleration of the car to the speed of 4 ms^{-1} (14 km/h) at negligible fuel consumption. Because of the fact that time of 2,0-2,5 s is spent for increase due to the additional fuel supply of rotational speed of the engine crankshaft to 200-210 s^{-1} and the corresponding increase in rotational speed of the driving wheels connected to its transmission (the clutch is in work!), then as the a result the car acceleration of at least 1,2 ms^{-2} and the speed of 7 ms^{-1} (25 km/h) is provided. At 180 such stages of working shift the bus will run $S_2 \approx (2\cdot 1{,}5+5{,}5 \cdot 2{,}5)\ 180 \text{ m} \approx 3$ km, having spent such fuel quantity which is given in the calculation. The average power necessary for driving wheels during the process of a speed set on the second gear at the same time will be

$$P_2 = m_{fm}\, g\, f\, V_{a2} + m_{fm}\, a_{x2} \delta_{f2}\, V_{a2} = 6415\cdot 9{,}81\cdot 0{,}012\cdot 5{,}5 + 6415\cdot 1{,}2\cdot 1{,}41\cdot 5{,}5 = 63{,}8 \text{ kW},$$

where g – acceleration of gravity; $f = 0{,}012$ – rolling resistance coefficient; V_{a2} – the average speed of the bus at acceleration on the 2nd gear; a_{x2} – acceleration on the 2nd gear; δ_{f2} – coefficient of accounting of the rotating masses on the 2nd gear.

Further at speed of 7 ms^{-1} there is a switching to the third gear (duration of switching t_s = 1 s), at the same time the way passed by the bus by inertia during a day will be defined by expression of $S_{j2} = 7 \bullet 180$ m ≈ 1,3 km.

The following stage is acceleration on the third gear. The bus engine and transmission provide acceleration a_{x3} = 0,8 ms^{-2}. Average speed at the stage V_{a3} equals 9,5 ms^{-1}, acceleration time to the speed of 12 ms^{-1} (≈ 43 km/h) t_{a3} = 6,25 s, therefore the way passed by bus on the 3rd gear during the working shift will make S_3 = 9,5•6,25•180 m = 10,7 km. So, the average power necessary for driving wheels on the third gear will be

$$P_3 = m_{fm}\, g\, f\, V_{a3} + m_{fm}\, a_{x3}\delta_{j3}V_{a3} + W_{sf}\,(V_{a3})^3 = 6415 \bullet 9{,}81 \bullet 0{,}012 \bullet 9{,}5 + 6415 \bullet 0{,}8 \bullet 1{,}16 \bullet 9{,}5 + 2{,}89 \bullet 9{,}5^3 = = 7174\ W + 56555\ W + 2479\ W = 66{,}2\ kW,$$

where W_{sf} – streamlining factor of the bus PAZ-3204 (W_{sf} = 2,89 $N.s^2m^{-2}$).

At first there is switching to the 4-th gear before the acceleration stage. With duration of process of switching about 1 sec the way passed by bus by inertia on this mode for daily working shift will be defined by expression of S_{j3} = 12•1•180 m ≈ 2,2 km. The bus engine and transmission provide on the 4th gear acceleration a_{x4} = 0,4 ms^{-2}. Average speed of the movement at this stage V_{a4}=14,35 ms^{-1}, acceleration time to the maximum speed of 16,7 ms^{-1} (60 km/h) at the specified acceleration t_{a4} = 11,75 s, therefore the way passed by the bus at acceleration on the 4th gear for working shift will make S_4 = 14,75•11,75•180 m = 30,35 km. Average power necessary for the leading wheels for acceleration on this gear

$$P_4 = m_{fm}\, g\, f\, V_{a4} + m_{fm}\, a_{x4}\, \delta_{j4}\, V_{a4} + W_{sf}\,(V_{a4})^3 = 6415 \bullet 9{,}81 \bullet 0{,}013 \bullet 14{,}75 + 6415 \bullet 0{,}4 \bullet 1{,}05 \bullet 14{,}75 + + 2{,}89 \bullet 14{,}75^3 = 11740\ W + 38670\ W + 8540\ W = 58{,}9\ кW.$$

Having reached maximum speed of 60 km/h permissible in the city, the bus has to continue to move on the fourth gear with constant speed. At the same time the necessity for the power driven to the leading wheels decrease significantly. It is defined by expression

$$P_{4const} = m_{fm}\, g\, f\, V_4^{max} + W_{sf}\,(V_4^{max})^3 = 6415 \bullet 9{,}81 \bullet 0{,}013 \bullet 16{,}7 + 2{,}89 \bullet 16{,}7^3 = 27{,}16\ кW.$$

The track length, passed by the bus during the working shift on the fourth gear with constant speed, represents the difference between the full daily run (240 km) and runs of S_2, S_3 and S_4 on three gears used for a speed set, runs of S_{j2} and S_{j3} by inertia at gear shifts and also S_{br} during the braking time by the engine about 100-120 m before each stop when fuel supply in cylinders of the engine is turned off. As a result it will make for the considered route

$$S_{4const} = 240 - S_{br} - S_{j2} - S_{j3} - S_2 - S_3 - S_4 = 240\ km - 18\ km - 1{,}3\ km - 2{,}2\ km - 3{,}0\ km - 10{,}7\ km - - 30{,}4\ km = 174{,}4\ km.$$

Fuel consumption [1, 5] is defined for all rated points of value **E** of ratio **E** = ω_e/ω_p and calculated extent of engine power use is **И** = $P_w / k_b\, \eta_t P_e^{вcx}$ according to the analytical method of definition of the wheel car. As previously reported we use ω_e and $P_e^{вcx}$, corresponding to average speed of the bus at the considered stage (see table 2). The obtained values **E** allow to calculate coefficient K_E by means of the expression given in literature [1, 5]. Respectively, the obtained values **И** allow to define $K_И$ coefficient by the adjusting polynom [4, 5] removed earlier. As a result there is a possibility of calculation of specific fuel consumption g_e for each specific mode of the bus movement on a route (table 2).

Table 2. Results of calculation of daily fuel consumption of the shuttle bus PAZ-3204, equipped with Cummins engine at the passable traffic circuit of 240 km of and 180 stops during the working shift

Parameter	Acceleration on the 2nd gear	Acceleration on the 3rd gear	Acceleration on the 4th gear	Motion on the 4th gear with V_{const} = 60 km/h
ω_{ea}, c^{-1}	165	167	165	192
E	0,61	0,615	0,61	0,71
K_E	0,956	0,956	0,956	0,955
И	0,88	0,91	0,775	0,34
$K_И$	0,924	0,937	0,931	1,059
g_e, g/кW.h	201,4	204,2	203,1	230,6

Q_{si}, l	2,3	6,0	9,5	24,5
Consumption, l/km	0,77	0,56	0,31	0,14
$Q_{S\Sigma}$, l		42,3		

The obtained g_e values allow to pass to calculation of fuel consumption by this bus at the movement on each of the considered stages. For this purpose it is possible to use expression (1) which is a little transformed in comparison with the initial one [1, 5]. For the considered situation it will be transformed in

$$Q_{si} = g_{ei}\,(P_{\psi i} + P_{wi})\,S_i\,/\,3600\;k_b\,\rho_t\,V\,\eta_t\,, \qquad (1)$$

where $S_i / 100$ is the relative size of the way passed at the corresponding stage of the movement of the bus during the working shift (12 trips along a route).

Apparently from table 2, rated sizes of fuel consumption at separate stages turned out as such: $Q_{s2} = 2{,}3$ l; $Q_{s3} = 6{,}0$ l; $Q_{s4} = 9{,}5$ l; $Q_{s4const} = 24{,}5$ l. Total daily fuel consumption by the bus on the specified route is made 42,3 l.

With observance of the technique algorithm given here calculation of daily fuel consumption of the moving bus PAZ-3204 on the same route and with respect to the same of acceleration dynamics at stages is made, but at 30 stops (360 stops for a full working shift) this situation is better that corresponds to the real situation of the movement in city streets. Fuel consumption on stages were distributed as follows: $Q_{s2} = 3{,}8$ l; $Q_{s3} = 11{,}8$ l; $Q_{s4} = 18{,}9$ l; $Q_{s4\text{пост}}$ equals 15,6 l. Total daily fuel consumption made 50,1 l.

We will compare the obtained rated results with the real operation data of this bus models. So, according to poll of drivers in the Nizhny Novgorod transport LLC "Football", the average value of fuel consumption at a daily run of 240 km on PAZ-3204 buses during the spring and autumn period of their operation is 45 ± 3 l. The obtained average calculated value (46,2 ± 3,9 l) gives an error of average size about 2,5% and this mistake takes into account the maximum dispersion of 11%. Therefore, the developed technique is quite applicable for practical purposes, in particular, for studying and the analysis of the specific operational problems connected with rationing of fuel consumption by wheel cars.

So, in actual practice drivers of shuttle buses are forced to stop at intersections and at traffic lights several times in addition to the stops planned on a route. The calculations which are carried out in this section show that each stop of the PAZ-3204 bus demands an additional expense about 0,03-0,05 l of fuel on the subsequent accelerations of the bus till the rated speed of the movement. Therefore for decrease in fuel consumption and increase in efficiency of city buses usage it is necessary to reduce quantity of unplanned stops. It depends to a large extent on the level of the organization of the movement in city streets and also on the ability of drivers to analyze the situation on the road quickly, in advance to foresee traffic light signals, and to able to apply actions for management of traffic flows like "green wave".

Also smoother accelerations of the bus from a stop with restriction of the maximum turns of the engine at level of 1700-1900 RPM and use of overdrive gear at the movement with constant speed till the rated speed of 50-60 km/h on direct gear is achieved though influence of these factors as calculations show, will affect fuel consumption reduction less significantly.

It should be noted that the important factor influencing fuel consumption is the maximum speed gathered by the city bus at the movement on a route. This conclusion is well confirmed by the example of calculation of daily fuel consumption for the bus PAZ-3204 used for public conveyances in Nizhny Novgorod as the share taxi route No. 40.

Route length is 24 km, it contains 44 planned stops. The analysis of traffic conditions on a route showed that in each run the driver is forced to make about 25-26 more necessary stops because of need of journey of a non-regulated intersection or a red signal of the traffic light. Therefore the total number of stops is calculated as equal 70. Thus at a daily run of 240 km the bus makes 10 runs, during the whole working day, i.e. the total number of stops for the working shift equals 700. At the same time the average size of a run between stops will be only 240/700 = 0,343 km. Therefore in most cases achievement of maximum speed of the movement of 60 km/h (16,7 ms^{-1}) which demands about 240-250 m for acceleration of the bus after a stop becomes not rational as right after it on the remained 100 meters it will be necessary to reduce the gained speed.

Observations showed that on a route No. 40 drivers are most often limited to a maximum speed set of 50-55 km/h (13,9-15,3 ms^{-1}). The average number of passengers inside the bus during the working shift insignificantly differs from the value accepted in the previous calculations - 19 people.

Table 3. Results of calculation of daily fuel consumption of the shuttle bus PAZ-3204, equipped with Cummins engine, at 240 km of the passable traffic circuit and 700 stops during the working shift

Parameter	Acceleration on the 2nd gear	Acceleration on the 3rd gear	Acceleration on the 4th gear	Motion on the 4th gear with $V_{constant}$ = 50 km/h
ω_{ecp}, c^{-1}	187	167	150	161
E	0,69	0,615	0,55	0,59
K_E	0,954	0,956	0,96	0,959
И	0,84	0,91	0,76	0,27
$K_И$	0,926	0,937	0,933	1,19
g_e, г/кW.h	201,6	203,8	204,2	260,2
Q_{si}, *l*	3,6	23,1	13,9	6,5
Consumption, *l*/km	0,79	0,56	0,31	0,14
$Q_{S\Sigma}$, *l*		47,1		

Results of table 3 correspond to the situation when the bus moved on the fourth (direct) gear with the maximum speed of 50 km/h. The possible option when the bus picked up maximum speed of 60 km/h for 14% of sites between the next stops gathered was also analyzed (taking into account the fact that real availability of several sites between stops is up to 1,0-1,5 km long). In this case total daily fuel consumption is made as 48,9 *l*.

On the analyzed route there is a sudden abrupt (α = 6%) rise which length is about 1 km. Calculation showed that taking into account that on the return run the bus moves already on descent, and the engine works as a motor without consumption of diesel fuel, this factor will affect a total daily expense slightly (increase in rated size of a daily fuel consumption is on 0,3-0,5 l).

In conclusion it is said that rated definition consideration of fuel consumption by wheel cars is necessary to pay attention to that fact that indicators of specific fuel consumptions on 1 km of the passable way (see table 2 and table 3) in both in detail considered rated cases were on the 3rd and 4th gears identical, and on the 2nd gear they differ for only 2,5%. It is clear that the rated modes of the movement of the bus were chosen close to optimum.

Comparison of the obtained rated data with normative indicators of fuel consumption by bus PAZ-3204 with the *Cummins B3.9 140* diesel engine in all considered cases appears not in favor of the Russian standard methodology of rationing of expenses of automobile kinds of fuel [6]. So, usually used when determining a standard consumption of fuel by the car without technological work (for buses – without fuel expenses on operation of the independent heater) the formula looks like:

$$Q_{SH} = 0{,}01\, H_S S\, [1 + 0{,}01(D)], \qquad (2)$$

where H_S – the basic (linear) consumption rate of fuel for the specific vehicle, *l*/100 km; S – vehicle mileage, km; D – the sum of the regulated extra charges for features of operation, age of the vehicle, an operation season, etc., %.

For Nizhny Novgorod city shuttle bus route No. 40, considered in this research work H_S = 23,7 *l*/100 km [6], S = 240 km, $D1$ = 25% (the movement in city streets with the population about 1,5 million people, in summertime, the bus passed a running in and its age is less than 5 years); $D2$ = 10% (frequent technology stops of the vehicle) [6]. Substitution of these values in formula (2) yields such result

$$Q_{SH} = 0{,}01 \cdot 23{,}7 \cdot 240\, [1 + 0{,}01 \cdot (25+10)] = 76{,}8\ l.$$

The received value demonstrates that the normative size of fuel consumption by the city shuttle bus (76,8 *l*) with daily run 240 km practically by one and a half times exceeds the value defined by the technique of rated determination of daily fuel consumption by the same bus presented in article (47,1+ + 0,5 = 47,6 *l* –

taking into account additional costs of overcoming kilometer rise, fivefold during working shift) at a similar run. It should be noted that similar overestimate is typical for any vehicle used in Russia. It is confirmed by the numerous special researches conducted at Motor transport department of NSTU n. a. R.E. Alekseev [7]. For example this fact is confirmed by the example of cargo road trains in the thesis was is successfully defended [8].

The specified overestimated expenses of automobile fuels by the standard rationing technique in Russia, their difference from real values significantly reduce volumes of the taxes to the budget as the considerable part of profit of the auto enterprises is taken away "in a shadow". Besides, conditions for artificial overestimate of cost value of automobile transportations, expenses on fuel as a part of which make at least 30%, are created.

Conclusion: the standard technique of fuel consumption rationing on the motor transport needs to be subjected to essential changing, and it has to be made it on the methodology basis of the researches provided in this article and in some other scientific researches [1, 4, 5, 8].

References

[1] Peskov V I and Kuzmin N A 2018 *Rated researches of utilization properties of cars* (Nizhny Novgorod: Nizhny Novgorod State Technical University n.a. R.E. Alekseev) p 212

[2] Moskvin R K and Kozyrev S V 2006 *Buses PAZ of new generation* (Moscow: Automotive industry No. 9) pp 19-21.

[3] *Rules No. 107. The uniform instructions concerning the official statement of vehicles of categories Sq.m and M3 concerning their general design* 2004 (Review 1. Addition 106/UNECE – Geneva)

[4] Peskov V I, Serdiuk V I and Serdiuk A E 2009 *Improvement of exploitation qualities of automobile* (Nizhny Novgorod: Nizhny Novgorod State Technical University n.a. R.E. Alekseev) p 135

[5] Kuzmin N A and Peskov V I 2017 *Theory of operational properties of the car* (Moscow: FORUM) p 256

[6] *Consumption rates of fuels and greases on the motor transport: Order of the Ministry of Transport of the Russian Federation* 2008 No. AM-23-r of 14.03.08. (Moscow) p 64

[7] Kuzmin N A 2014 *Technical operation of cars: rationing and management* (Moscow: FORUM) p 224

[8] Borisov G V, Kuzmin N A and Erofeyeva L N 2015 *Analytical approach to rationing of a consumption of automobile fuels* (Moscow: Intelligence. Innovations. Investments. No. 3) pp 91-96.

IOP Conf. Series: Materials Science and Engineering **386** (2018) 012023 doi:10.1088/1757-899X/386/1/012023

Research of composite coverings wear resistance for forestry machines by electro contact sintering

V M Ged'o [1], V A Markov [2], V A Sokolova [3], G K Parfenopolo [4], L G Chernyh [5]

[1] Candidate of Technical Sciences (PhD)

[2] Candidate of Technical Sciences (PhD), Assistant Professor

[3] Candidate of Technical Sciences (PhD), Assistant Professor

[4] Senior lecturer, Assistant Professor

[5] postgraduate student

St. Petersburg State Forest Technical University

mactor85@mail.ru

Abstract. Many parts of the chassis in forestry machines are exposed to heavy wear due to operation under heavy-duty conditions. Parts exposed to the heaviest wear are friction bearings and axles; whereas it is often cheaper to replace bearings than to restore them, axles might be quite expensive. This study examines a promising method of axle-class parts, improving their wear resistance. Analysis of the study results allows to come to a conclusion that coating produced by electric-contact sintering of composite materials featured a > 1.5x wear resistance of 40Kh steel from which the representative roller axle was made. Therefore, the service life and reliability of joints made of or restored by means of composite materials is considerably better than that of the standard forestry-machine chassis connections.

1. Introduction

Many parts of the chassis in forestry machines are exposed to heavy wear due to operation under heavy-duty conditions. Parts exposed to the heaviest wear are friction bearings and axles; whereas it is often cheaper to replace bearings than to restore them, axles might be quite expensive. Abrasive friction is the cause of wear of axles in the forestry-machine chassis. Friction is a very common phenomenon. It occurs in every relative motion of contiguous bodies or parts thereof. The roller axle of an Onezhets-300 skidder (Fig. 1) was chosen as the representative part for studying wear resistance.

Electric-contact sintering of composite material can prolong the service life of roller axles while strengthening their operating surfaces; as such, it is proposed as a method for roller pin restoration. One fact that must be born in mind is that implementing an electric-contact sintering unit is not associated with considerable financial investments and is affordable for most enterprises in the industry.

This study examines the wear resistance of parts restored by this technology under close-to-real operating conditions. Special attention is paid to analyzing the effect of different sintering parameters (current, voltage, and electrode pressure force) on the wear resistance of restored parts.

IOP Conf. Series: Materials Science and Engineering **386** (2018) 012023 doi:10.1088/1757-899X/386/1/012023

Figure 1.Skidder roller axle

2. Research methodology

Figure 2 demonstrates technology of electric-contact sintering of metal powders unto part surfaces. The technology shown therein is intended for the restoration of cylindrical parts. Current is applied to the welding zone via two welding rollers. This eliminates contact transitions from the outer loop, thus considerably reducing the power loss.

The process of electric-contact welding is essential concerning sintering a layer of metal powder onto the part surface pressurized by the forming roller and exposed to high-density current applied to the powder layer (3000 to 5000 A/cm2). That is, the powder is sintered by the heat generated by the electric current against active resistance [3].

Friction is referred to as external friction if it occurs in the relative displacement of two contiguous bodies. It depends on the external superficial interaction of these bodies near the point of contact, but not upon inner conditions. External friction and wear is a three-stage process: first the surfaces interact, then the superficial-layer material is altered by friction, then the surfaces are destroyed. In this case, the roller axle is exposed to friction against a harder part, a roughness 15 (pure bearing steel) steel bearing. Surfaces of solids feature micro relief. Micro relief in its turn features such attributes as shape deviation, waviness, and roughness. One can conclude that friction-caused structural alteration occurs at certain points of contact. This is why the structure and properties of the superficial layer are different from the structure and properties characteristic of volumetric deformations and heat treatment.

1. the part to be restored; 2. the weld metal;
3. the welding roller; 4. metal powder; 5. the bunker;
6 . the strengthening roller.
Figure 2. Electric-contact sintering of powder materials.

In this case, the roughness of both parts equals to Ra 2.5, which minimizes the effect of this factor. When considering external friction, three factor groups must be borne in mind, see Table 1 [1].

Table 1

Factors affecting the external friction of solids.

Input factors	Internal factors	Output factors
Nature of solids exposed to friction Lubricant Load Velocity	Alteration of roughness Alteration of properties of superficial films Heat output Alteration of structure	Friction force Wear intensity

IOP Conf. Series: Materials Science and Engineering **386** (2018) 012023 doi:10.1088/1757-899X/386/1/012023

Temperature	Alteration of mechanical properties	

To reproduce natural friction conditions, we used a DM-15 machine, i.e. a dynamic machine with a 15-kN maximum load of the tested joint.

This machine is able to reproduce abrasive friction by applying such abrasive elements as sand, soil, etc. to the part contact point. The real roller-axle to inner bearing race joint was used as the friction couple. Before testing, both parts were lubricated with UNIOL-1 (GOST 1033-79).

The parts were marked at the measurement points. The conjugate diameters were measured post-disassembly by means of an electronic micrometer and a caliper with an accuracy of 0.001 mm; measurements were done in two perpendicular- planes and three points (center and edge points), whereby wear was assumed to be the average value of those points.

The load upon the conjugate parts was assumed on the basis of the real value P=15000 N as measured by a DOSM-5-1 (Rus: ДОСМ-5-1) dynamometer, TU25.06.629-74. The following requirements: were also met: rotary motion, maximum sliding velocity of the friction couple Vmax=0.07 m/s at the output shaft rotation speed n=30 min- as transmitted to it by the machine's motor. Prior to experimentation, each friction couple was broken in at P=5000 N and V=0.01 m/s. During the experiments, the temperature in the friction zone was measured by means of a thermometer placed in a hole in the bearing race.

To evaluate the wear rate, the following dependencies were assumed [4]:

$$I = \frac{U}{t} \quad (1)$$

where U is the wear of the conjugate parts, mm;

t is the machine runtime, h.

$$I_{abrasion} = I_{spindle} + I_{bearing} \qquad (2)$$

where $I_{abrasion}$ is the total connection wear rate, mm/h;

I_{axle} is the axle wear rate, mm/h;

$I_{bearing}$ is the bearing wear rate, mm/h.

Assume: the uncoated roller-axle wear rate to be the nominal wear rate ($I_{nominal}$), then by preliminary experimentation, we find that $I_{nominal}$ = 0.014 mm/h. Nominal wear resistance was studied under the same conditions as the wear resistance of coated axles.

The following factors of the composite-material sintering process were assumed to have an effect on the wear intensity: Current (A), Voltage (U), and electrode pressure force (P). the factor levels and their variation intervals are given in Table 2..

Table 2

Factor levels

Factor level	I, kA		U, V		P, kN	
	x1	lnx1	x2	lnx2	x3	lnx3
Lower (-1)	10	2.30	2	0.69	0.5	-0.69
Basic (0)	12	-	3	-	1	-
Upper (+1)	14	2.64	4	1.39	1.5	0.4
-α	7.85	-	1.57	-	0.285	-
+α	17.01	-	4.86	-	1.715	-

Using Prof. Yu. M. Zubarev's method [2], one can find from the table:

$$X1=2\times(\ln x1-2.3)/[2.3-2.64]+1=-5.88\ln x1+14.5$$

$$X2=2\times(\ln x2-0.69)/[0.69-1.39]+1=-2.86\ln x2+5.83$$

$$X3=2\times(\ln x3+0.69)/[-0.69-0{,}4]+1=-0.96\ln x3-0.63$$

We further performed regressive analysis. The experimental results are encoded and summarized in Table 3. For greater confidence, we added 7 support experiments to the 8 main experiments. [2]

Table 3

Experiment planning matrix

Experiment #	X1 code	X2 code	X3 code	For r1 I', mm/h	For r2 I'', mm/h	$\bar{I}$, mm/hour	$\bar{Y} = ln\bar{I}$
1	+	+	+	0.008	0.009	0.0085	-4.77
2	-	+	+	0.015	0.017	0.016	-4.14
3	+	-	+	0.014	0.012	0.013	-4.34
4	-	-	+	0.019	0.021	0.02	-3.91
5	+	+	-	0.009	0.010	0.0095	-4.66
6	-	+	-	0.016	0.018	0.017	-4.07
7	+	-	-	0.015	0.017	0.016	-4.14
8	-	-	-	0.021	0.023	0.022	-3.82
9	-α	0	0	0.024	0.025	0.0245	-3.71
10	+α	0	0	0.013	0.014	0.0135	-4.31
11	0	-α	0	0.019	0.019	0.019	-3.96
12	0	+α	0	0.015	0.015	0.015	-4.20

Table 3: Continuation

13	0	0	-α	0.019	0.019	0.019	-3.96
14	0	0	+α	0.018	0.017	0.0175	-4.05
15	0	0	0	0.016	0.018	0.017	-4.07

Let us further write an equation introducing the terms that take into account the factor interaction, and represent it in encoded variables:

$$Y = \tau = b_o + \sum_{i=1}^{3} b_i x_i + \sum_{i,j} b_{ij} x_i x_j + b_{123} x_1 x_2 x_3 + \sum_{i=1}^{3} b_i x_i^2 \quad (3)$$

Regression-equation coefficients are found as follows:

$$b_i = \partial Y / \partial X_i$$

By substituting bi in equation 3, we obtain:

$$Y=-33.85-1.97X_1-1.43X_2-0.47X_3-0.47X_1X_2-0.15X_1X_3+0.11X_2X_3+0.07X_1X_2X_3-0.06X_{12}-0.24X_{22}-0.09X_{32} \quad (4)$$

Formula 5 is used to find the experiment reproducibility variance, and the calculated values are summarized in Table 4.

Experiment reproducibility variance:

$$S_y^2 = \frac{\sum\left(\bar{Y}_u - Y_{uq}\right)^2}{N-(r-1)} \quad (5)$$

where Yuq is the value of Yu in the qth repeated experiment.

Table 4.

Experiment reproducibility variance

Experiment #	$\bar{Y}_u - Y_{uq}$	$(\bar{Y}_u - Y_{uq})^2$	Experiment #	$\bar{Y}_u - Y_{uq}$	$(\bar{Y}_u - Y_{uq})^2$
1	-4.77+4.83=0.06	0.0036	5	-4.66+4.71=0.05	0.0025
1	-4.77+4.71=-0.06	0.0036	5	-4.66+4.6=-0.06	0.0036
2	-4.14+4.2=0.06	0.0036	6	-4.07+4.14=0.07	0.0049

2	-4.14+4.07=-0.07	0.0049	6	-4.07+4.02=-0.05	0.0025
3	-4.34+4.27=-0.07	0.0049	7	-4.14+4.2=0.06	0.0036
3	-4.34+4.42=0.08	0.0064	7	-4.14+4.07=-0.07	0.0049
4	-3.91+3.96=0.05	0.0025	8	-3.82+3.86=0.04	0.0016
4	-3.91+3.86=-0.05	0.0025	8	-3.82+3.77=-0.05	0.0025

Sy2=0,0581/8· (2-1)=0,00726

Sy=0,085

Variance homogeneity is verified by Cochran's test:

$$G^{расч} = \frac{S^2_{yu\max}}{\sum S^2_{yu}} = \frac{0,0064}{0,0581} = 0,11$$

For N=8; α=0.05; f=r-1=1 the critical Cochran value is found from the table: GKP=0.6798.

As $G_{calculating}$ < GKP, accept the hypothesis.

We further verify significance of the regression coefficients:

Sbi2=Sy2/(N×r)=0.00726/(8· 2)=0.00036; Sbi=0021

The critical test value is determined from the table: tKP=2.306.

The half-length of the confidence interval is determined by:

$$\Delta b_i = t_{KP} \cdot S_{bi} = 22,327 \cdot 0,019 = 0,42$$

The coefficient is significant if |bi |≥Δbi. By removing the insignificant coefficients the regression equation is written as:

$$Y=-33.85-1.97X_1-1.43X_2-0.47X_3-0.47X_1X_2 \qquad (5)$$

Model adequacy verification. The adequacy variance is:

$$S^2_{АД} = 2/(8-5) \cdot 0,022 = 0,0147$$

The Fisher's test value can be found as :

$$F = S^2_{АД} / S^2_y = 0,12 / 0,00726 = 16,5$$

If in the table FKP=19.37 > 16.5, therefore, the model is adequate.

3. Research results

By transition to natural variables and equation involution, we find:

$$I=0.0455-0.00175I-0.0025U-0.00175P \qquad (6)$$

From this equation, one can conclude that all the factors of the process are significant.

Let us further optimize it by the Monte-Carlo method using SPSS v.13.

The process parameters: I = 14 kA, U =4V и P=0.5 kN at I_{m}=0.009 mm/ha are found by optimization.

Given that $I_{nominal}$ = 0.014 mm/h, we find the wear coefficient:

$$K_{degree\_of\_wear.} = \frac{I_{nominal}}{I_{experimental}} = \frac{0,014}{0,009} = 1,55 \qquad (7)$$

4. 4. Conclusions

Analysis of the study results allows to make a conclusion that coating produced by electric-contact sintering of composite materials featured a > 1.5x wear resistance of 40Kh steel from which the representative roller axle was made. Therefore, the service life and reliability of joints made of, or restored by means of, composite materials are considerably better than that of the standard forestry-machine chassis connections.

References:

[1] Gerasimov. S. A. Jour. Friction in Movable Joints (Treniye v podvizhnykh soyedineniyakh). Nauka i obrazovaniye, No. FS77, 2004, P. 34-42.

[2] Zubarev Yu. M. Using the Methods of the Multi-Factorial Experiment Planning in Mechanical Engineering (Primeneniye metodov teorii planirovaniya mnogofaktornykh eksperimentov v tekhnologii mashinostroyeniya). St. Petersburg: OP PIMash, 2000, 130 p.

[3] Klimenko Yu.V. Electric-Contact Welding (Elektrokontaktnaya naplavka). Moscow: Metallurgiya, 1978, 127 p.

[4] Pekarev. A. R, Pekarev. A. A, Restoration of Movable Joints (Vosstanovleniye podvizhnykh sopryazheny) — Cheboksary: ServisMash, 1985 – 80 p.

[5] Anisimov G. M. Main Areas of Improving the Operating Efficiency of Crawler Skidders (Osnovnye napravleniya povysheniya ekspluatatsionnoy effektivnosti gusenichnykh trelevochnykh traktorov) — St. Petersburg: SPU, 2007. - 456 p.

IOP Conf. Series: Materials Science and Engineering **386** (2018) 012024 doi:10.1088/1757-899X/386/1/012024

Study of factors affecting the reliability of turbochargers

L A Berdnikov, A A Pikulkin, M G Korchazhkin, P I Bazhan, L A Zakharov

Nizhny Novgorod state technical university named after R.Y. Alekseev (NSTU)

E-mail: kormg@list.ru

Abstract. The thermal loading factors of the turbocharger affecting the reliability are considered. The ways of reducing the heat load of the turbocharger elements are analyzed. The scheme of the turbocharger rotor thermal optimization, as well as options for changing the shape of the rotor. The thermal analysis of the rotor by the finite element method is performed. The temperature distribution on the surface and in the rotor section is obtained. Composed vector representation of the heat flux at the nodes of the rotor. The conclusion about the influence of changes in the shape of the turbocharger rotor on the distribution of thermal energy is given.

Turbocharger (TCR) can be attributed to those components of the car, the reliability of which is strongly determined by the operating conditions of the vehicle and maintenance. Therefore, along with the workmanship of TCR and the correct selection of the corresponding internal combustion engine, a crucial role in the extension of its service life plays the correct operation and execution. The ultimate goal of all measures taken to improve the reliability can be considered to obtain the values of the life of the TCR close to the life of the engine as a whole. Point 4.7 GOST R 53637-2009 "Turbochargers automotive" establishes the following requirement for turbochargers: the resource of the turbocharger (its service life before overhaul) should be at least the corresponding indicator of the engine.

The most important factor affecting the reliability of the transport means, regardless of the applied scheme of the installation is the thermal loading of its parts. Due to the high temperature, such failures as leaks and coking of oil, jamming and overheating of the rotor, bearing unit are common. The key places of occurrence of negative consequences of high temperatures are the labyrinth ring seals of the rotor from the turbine wheel, as well as the bearing unit. In the first case, there is the problem of separating the hot gas medium and the lubricant medium (similar to the problem of separating the combustion chamber and the crankcase space in the ice in the form of coking piston rings), and the rotor rotation speeds practically do not allow the use of other types of seals. The temperature in the area of the labyrinth seal is higher than in the bearing unit, and the oil that got into the seal does not participate in the circulation in the lubrication system of the TCR and the ice and, as a result, does not remove heat from this unit, acquiring the temperature of the surrounding parts. In the case of the bearing, there is a problem of loss of lubricating properties of the oil when exceeding the permissible temperature values in friction pairs, with further heating, wear, failure.

Oil contact with parts having a high temperature, inevitably leads to an acceleration of its aging processes. Therefore, regardless of the issue of maintaining the efficiency of the TCR, it should strive for an initial approximation of the temperature of the walls of the cavities in the TCR housing to the temperature of the oil entering the TCR (90 – 110 ° C).

There are two main ways of heat energy distribution in TCR:

- from the exhaust gases to the turbine part body and to the Central body;
- from the exhaust gases to the turbine wheel and further on the rotor.

Methods of reducing the temperature in the above parts of the TCR are also two: thermal insulation and cooling by means of heat removal.

Thermal insulation involves reducing the thermal conductivity of materials λ in the heat flow path. In addition, it is possible to reduce the value of the total heat flow by reducing the cross section of the part in the direction normal to the direction of the main heat flow.

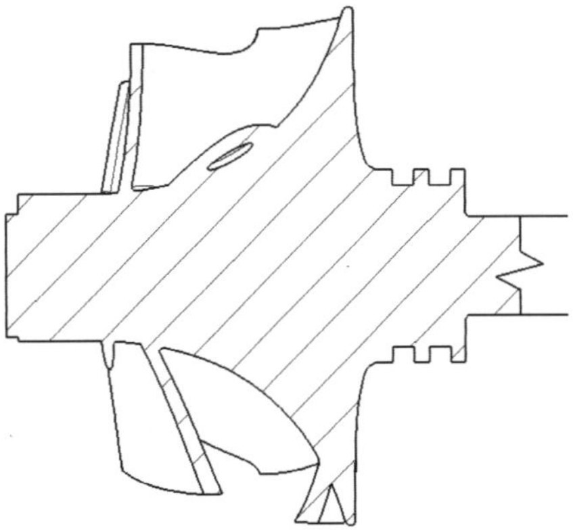

Fig. 1 Local section of the TCR rotor .

As can be seen from the figure 1, the rotor shaft in the region of grooves for labyrinth seal rings and the key points 1 and 2 considered in Fig. 2 are in close proximity to the impeller of the turbine.

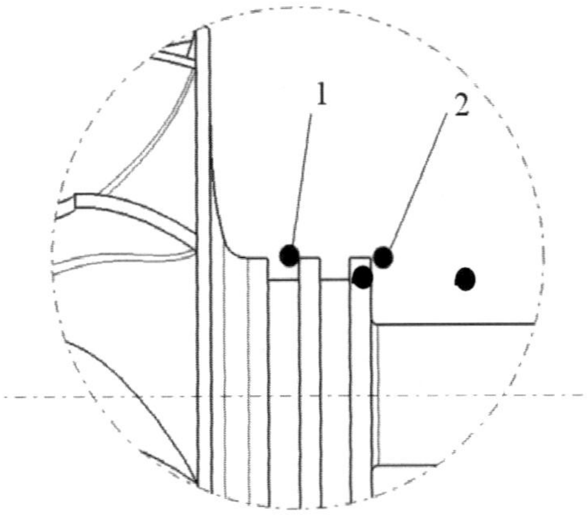

Fig. 2 Key points of the turbocharger rotor.

The latter, in turn, has direct contact with the high – speed flow of exhaust gases having a temperature of up to 700-1200 ° C depending on the type of engine. All this can cause the temperature in the areas under consideration to rise to critical temperatures in terms of oil properties. Variants of changes in the geometry of the rotor suggest the removal of heat flow from the key areas and, as a result, a decrease in the values of their temperatures.

In general, the optimization of parts under specified conditions, taking into account manufacturing techniques can be carried out by changing its geometry, changing materials (hence their properties) and a combination of these methods.

Of course, the consideration of the thermal problem for parts that experience high dynamic complex loads, in particular for the TCR rotor, is impossible without taking into account the changes in strength characteristics that change after the modification of the geometry. Therefore, it is necessary to carry out a number of strength analyses to verify the preservation of the rotor. Thus, the modification and complex analysis involves the implementation of the following scheme:

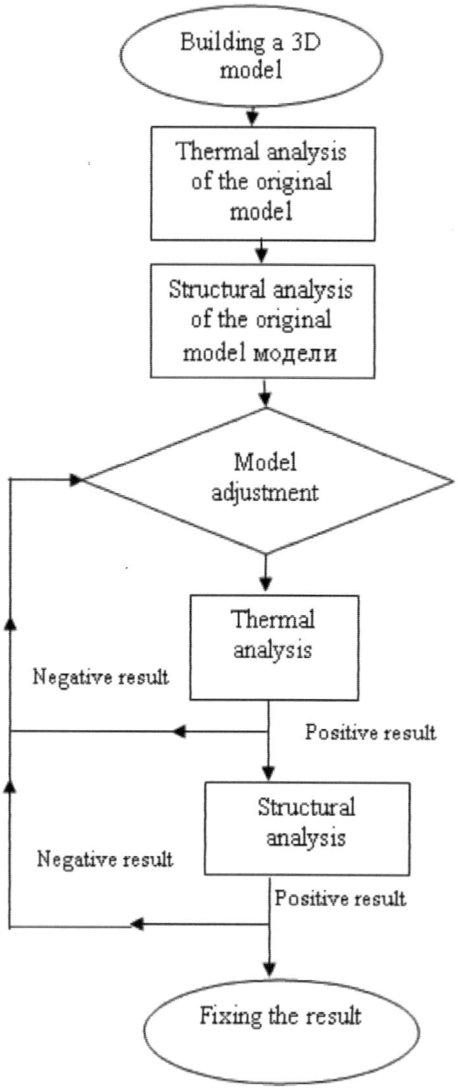

Fig.3 Simplified diagram of a thermal optimization.

A comparison method was used to assess the effectiveness of the changes. In the conducted analyses for the initial variant of the rotor and its modifications identical initial conditions, equal loads and identical structural and thermal physical characteristics of materials were set. This method allows to exclude errors when setting initial values (boundary conditions, etc.), since only the relative change of the TCR rotor parameters of interest is estimated. Despite this, the input numerical values of the initial parameters are close to the real ones, so the data obtained during the analysis can be used as indicative values.

Geometric changes of the rotor and its mass-inertia characteristics as a reduction factor

In accordance with the experience of production and operation, the shaft diameter of most turbochargers is 0.15 – 0.17 from the diameter of the compressor wheel, while there is a tendency to reduce it, because in this case, the efficiency of the TCR is increased by increasing the mechanical efficiency. The limit of shaft diameter reduction is determined mainly by technological difficulties, and not by considerations of design or operational nature, since the mechanical loads perceived by the shaft are small compared to its strength for bending and torsion. So the rotor of the turbocharger. Installed on a diesel engine with a power of 240 kW, rotates at a frequency of 60,000 min-1 and transmits a torque of about 0.5 Nm, the centrifugal forces from the residual imbalance of 0.15 g·cm do not exceed 6 N. There are already completed designs with a relative shaft diameter of 0.12, which show good reliability and durability. We present some possible variations in the rotor from a primitive rectangular groove to ducts with more complex geometry:

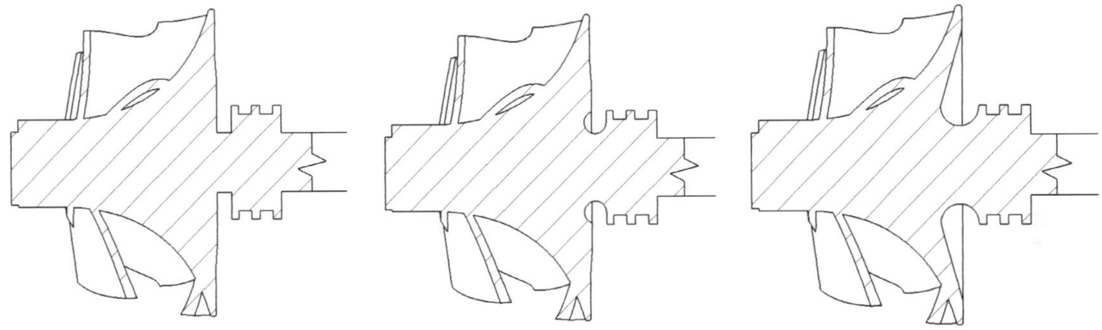

a) b) c)

Fig. 4 Options for changing the shape of the TCR rotor.

It should be noted that the reduction of the absolute values of the size of these ducts does not make sense. This would be useful if the solution provided for a particular model of TCR were to be specified.

The most obvious impact of these changes in the form have on the mass-inertia characteristics. CATIA software allows you to quickly determine the values of these indicators when making adjustments to the original solid model.

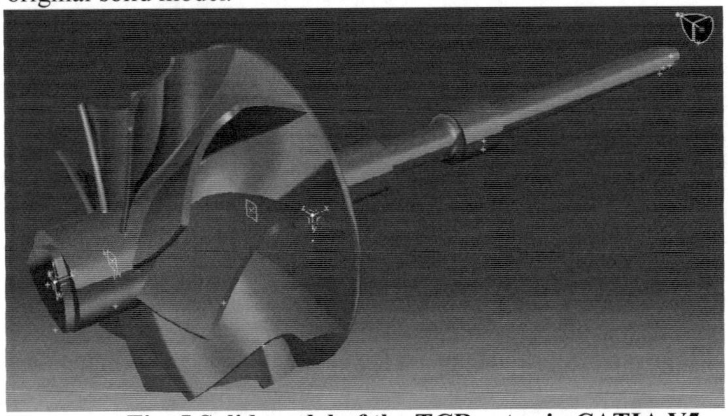

Fig. 5 Solid model of the TCR rotor in CATIA V5.

The main mass-inertial characteristics of the rotor variants at the accepted density of 7860 kg / m^3 and their relative change are summarized in **Table 1**

Table 1. Changes in the mass-inertial characteristics of the rotor

Parameters / Variant	Weight, kg	Relative change, %	Axial moment of inertia, kg·mm^2	Relative change, %
«0»	0,106	0	9,069	0
«а»	0,104	- 1,89	8,955	- 1,26
«б»	0,103	- 2,83	8,925	- 1,59
«в»	0,094	- 11,32	7,855	- 13,39

Thus, a significant change in the values of mass and moment of inertia takes place for the "в" variant. The reduction of the rotor moment of inertia has a positive effect on the turbocharger performance. The light rotor quickly increases its speed with increasing load on the engine, which increases the convenience of driving and improves the environmental performance of the engine (mainly diesel) in transient conditions by reducing the likelihood of lack of air with a sharp increase in fuel supply. Among other things, the decrease in the mass parameters of the rotor causes a decrease in the reactions of the bearing unit, for example, in the manifestation of the gyroscopic effect, which reduces the rate of wear of friction pairs.

Thermal analysis of the TCR rotor

The main purpose of the changes was to optimize the thermal condition of the rotor TCR, namely the reduction of temperatures in the zone of labyrinth seals and bearings. To analyze the effectiveness of the measures taken, a thermal analysis of the TCR rotor model was carried out using a software complex for the implementation of the ANSYS finite element method.

Stationary thermal analysis determines the steady-state temperature distribution in the structure and conductive heat flows. It is possible to set such “loads " as convective heat transfer from the surface, heat fluxes, heat flux density, heat source power and set temperatures. The analysis can be linear or nonlinear.

In the linear steady-state heat transfer process there is no influence of “thermal” masses (specific heat) and does not take into account the dependence of the thermal properties of the material on temperature. The derivative of temperature with respect to time $\{T\}$ is equal to zero, and the matrix coefficients of effective thermal conductivity is constant. In this case, the resolving equation is given by the form:

$$K] \{T\} = \{Q\}. \quad (1)$$

This system of linear joint equations is solved in one iteration and is used to calculate the processes of conductive and linear convective heat transfer.

The nonlinear non-stationary analysis of heat transfer does not consider time-dependent effects (there are no “thermal” masses). However, the thermophysical properties of materials (including the convective heat transfer coefficient from the surface) may vary with temperature; in addition, radiant heat transfer may occur.

The mechanism of radiative heat transfer is described in three different ways. A linear radiation finite element is used to model the transfer of heat by radiation between two points of space. The finite element of surface radiation is used to describe the radiation heat transfer between the surface and the point. The matrix generator is used to solve problems related to several absorbing and radiating surfaces. In the latter case, it is possible to take into account the complete or partial overlap of the surfaces, as well as to specify a node in space that absorbs or radiates energy. In the General case of heat transfer by radiation, the radiation heat flux density $\{Q\}$is a function of T^4 rather than T, i.e. the nature of the process is clearly nonlinear.

In nonlinear analysis, the thermal conductivity matrix is a function of temperature, and the solution of the problem is achieved by using iterations. In the program ANSYS, the iterative procedure has its main Newton-Raphson method, which involves solving a sequence of linear problems to obtain a nonlinear approximation. Thus, the equation for the nonlinear stationary heat conduction problem has the form

$$[K]_i \{\Delta T\}_{i+1} = \{Q^A\} - \{Q^{NR}\}_i , \quad (2)$$

where i is the iteration number.

This equation is solved in the first iteration at some initial temperature (which can be set by the user); in subsequent iterations, the temperature values obtained in the previous iteration are used to calculate the thermal conductivity matrix coefficients. The iteration process continues until the convergence of the solution is reached, i.e. until the user-defined convergence criterion is met. The convergence is controlled by the magnitude of the load vector residual (heat flow) and/or temperature change from iteration to iteration.

The solution results, for linear and nonlinear analysis, represent the temperature and heat flux density values at the nodes. These data can be used in post-processing to build a picture of isotherms in the computational model. Postprocessor tools can be used to obtain such specific information as the values of temperature gradients or flows in the nodes and in the center of the element, as well as the density of the heat flux through the radiation (absorption) surface. The information is displayed in tabular or graphical form.

Some assumptions were made in this analysis:

- there is only convective heat transfer, other types of heat transfer are taken into account by the heat transfer coefficient ; thus, the initial conditions are the boundary conditions of the third kind;
- a simplified model of the rotor (without impeller blades) is considered, which does not affect the result in the form of a picture of temperature fields;
- preliminary calculations revealed the absence of heat exchange effect in the part of the rotor shaft, which has contact with the compressor wheel and the air inlet flow; further calculations are carried out with the simplification of convective heat transfer in this part in the form of setting the same boundary conditions as the bearing part of the rotor;
- heat generation in friction pairs is negligibly low.

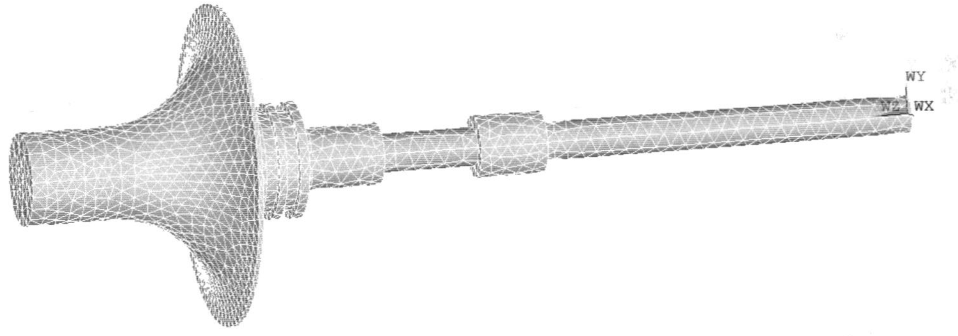

Fig. 6 Simplified solid rotor model with finite element mesh.

The analysis uses an arbitrary grid of finite elements-tetrahedra with a linear size of parabolic edges up to 1.5 mm, which provides the necessary accuracy of the solution at an acceptable computation time.

The analysis of any type requires specifying the properties of the material. The model used data tables for two materials-heat-resistant Nickel alloy turbine wheel and carbon steel shaft. The software package ANSYS provides the ability to take into account the dependence of the properties of materials on temperature. The properties of the materials are isotropic. Data on the characteristics of the materials are presented in table 2

IOP Conf. Series: Materials Science and Engineering **386** (2018) 012024 doi:10.1088/1757-899X/386/1/012024

Table 2. Determination of material properties.

Characteristic for materials I and II \ Temperature: °C		20	100	200	300	400	500	600
Density, kg/m^3	I	8450	8442	8431	8420	8408	8395	8383
	II	7820	7800	7770	7740	7700	7670	7630
Specific heat, j/(kg ·°C)	I	538	575	609	634	676	735	782
	II	472	482	496	513	532	555	583
Thermal conductivity, $\frac{Vt}{m \cdot {}^{\circ}C}$	I	7,9	8,8	10,5	11,8	13,8	15,9	18,4
	II	53	50,2	46	41,9	37,7	33,5	29,3
The modulus of elasticity, hPa	I	210	206	202	198	193	188	182
	II	215	210	205	200	190	180	170
Expansion coefficient $\alpha \cdot 10^5$ 1/°C	I	1,2						
	II	1,3						

I - heat-resistant nickel alloy, II - steel

Boundary conditions in the analysis are set for surfaces and include the temperature of the medium (liquid or gaseous) T_f washing the surface and the generalizing heat transfer coefficient α, taking into account both the nature of the medium flow around the surface and its thermal properties. The boundary conditions for thermal analysis are shown in figure 3.7:

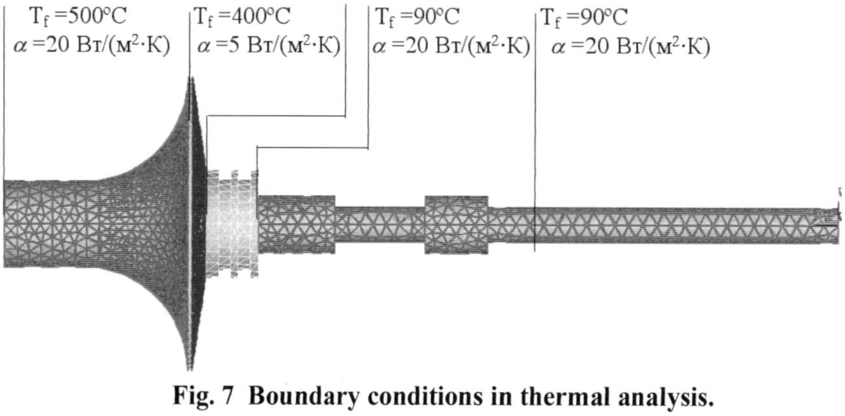

Fig. 7 Boundary conditions in thermal analysis.

Thus, a solid model with finite element partitioning, given material properties and boundary conditions form the basis for the implementation of the finite element method algorithm and obtaining the required results.

The main purpose of the analysis is to obtain the temperature distribution and the nature of the flow of heat in the rotor body for different variants of its shape. Changes in these parameters when modifying the geometry under similar initial conditions allow to draw conclusions about the impact of these changes in geometry on the thermal state of the rotor. We present the results of stationary thermal analysis of rotor variants starting from the initial modification. In figures (8-9) - the temperature distribution on the surface and in the rotor section, as well as the vector representation of the heat flow in the element nodes.

IOP Conf. Series: Materials Science and Engineering **386** (2018) 012024 doi:10.1088/1757-899X/386/1/012024

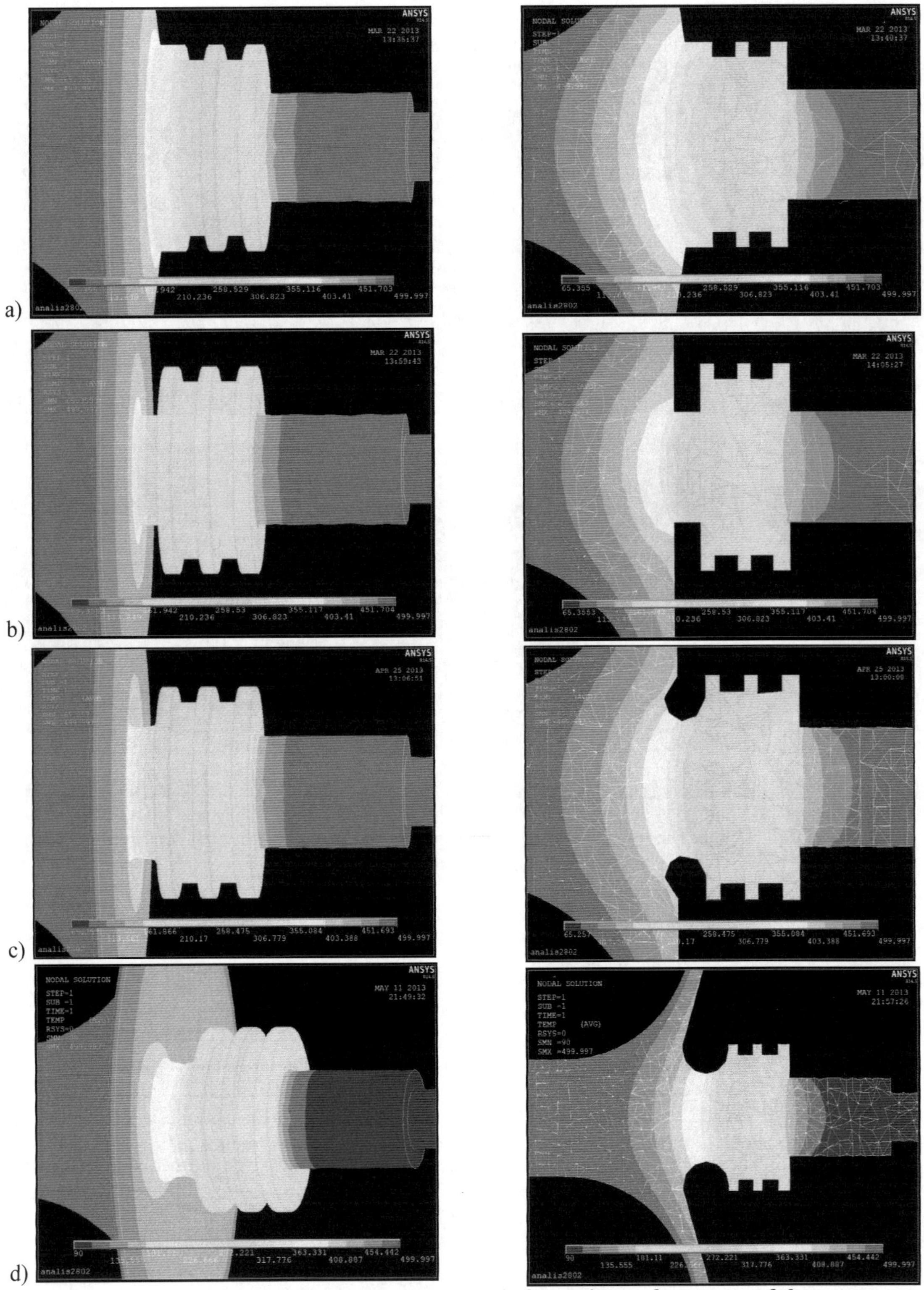

Fig. 8 The results of thermal analysis for variants of geometry of the rotor.

To facilitate the representation of the nature of the heat flow in the rotor body, let us present its vector visualization. The results are presented with different densities of vectors.

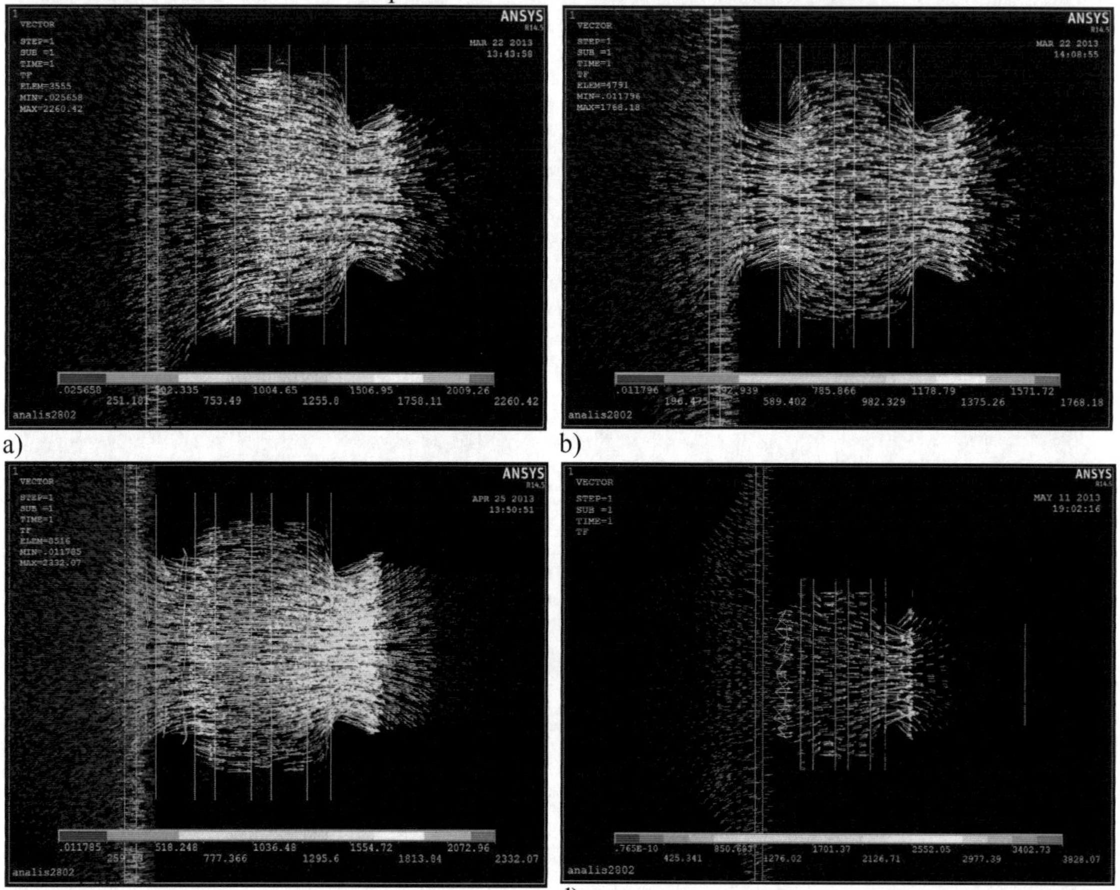

a) b) c) d)

Fig. 9 Vector representation of heat fluxes in the rotor body (with different vector densities).

The results of the temperature ranges for key points 1 and 2 are summarized in table 3

Table 3. Thermal analysis results for rotor shape options

Zone	Variant:	0)	A)	Б)	В)
T.1	Temperature, °C	306 - 330	210 - 234	258 - 282	272 - 294
	Relative change, %	0	- 30,2	- 15,1	- 11,0
T.2	Temperature, °C	210 - 234	185 - 210	210 - 234	204 - 226
	Relative change, %	0	- 11,0	0	- 3,1

The results suggest that the most effective way to reduce the temperature in the key areas is option (a), which can be explained by the smallest cross section in the heat flow path. Options б) and в) are also effective and contribute to a significant reduction in heat flow from the turbine wheel to the rotor shaft.

Thus, the shape change has a significant impact on the distribution of heat energy, allowing to reduce the temperature in the areas of interest. The ability to reduce temperatures by changing the shape of parts is limited by the strength characteristics. To assess the influence of the part shape on its structural qualities, it is necessary to conduct a number of subsequent analyses.

References

[1] Berdnikov, L. A., Suvorov I. A., Vilkov, N. A.. An integrated approach to the consideration of the factors determining the reliability of turbochargers of automobile ice. Collection of materials 12 International youth scientific and technical conference "the Future of technical science", Nizhny Novgorod NSTU 2013- p. 195-196

[2] Berdnikov, L. A., Suvorov I. A. "Study of the possibility of thermal optimization of the turbocharger rotor with the conduct of finite element analysis. The works of NSTU No. 4, 2013 p. 56-65»

[3] Berdnikov, L. A., Suvorov I. A. "Improvement of temperature condition of automobile engine turbochargers." Moscow. Motor transport enterprise. No. 4. April 2014-p. 40-42

[4] Berdnikov, L. A., Suvorov I. A., Kuzmin N. A., Thermal insulation screen of the turbocharger. Patent № 141164, Russia 22.04.2014

[5] Berdnikov, L. A., Suvorov I. A., The rotor shaft of the turbocharger. Russia, from 01.06.2015

[6] Berdnikov, L. A., Suvorov I. A., Design to prevent oil from burning in the turbocharger. Patent No. 159068. Russia 17.06.2015

Theoretical research of soil packing by timber harvester running gear

E E Klubnichkin, V E Klubnichkin, G O Kotiev

Bauman Moscow State Technical University, 5/1, 2-ya Baumanskaya street, Moscow, 105005, Russia

E-mail: vklubnichkin@gmail.com

Abstract. The problem of ecological compatibility of self-propelled vehicles and soils is intricate and requires integrated systematic analysis of multiple tasks. This work is aimed at studying theoretical researches of soil packing caused by timber harvester running gear. To preserve and ensure faster restoration of forest soil fertility it is necessary to define at the cutting area development stage the tolerable limit of the timber harvesting equipment's impact at which soil disturbance does not become irreversible and the process of restoration of its properties is the shortest.

1. Introduction

Machinery impact on forest growth conditions and forest regeneration have been studied from the beginning of mechanization of logging operations up until recently. In due time those issues have been examined by Melekhov [1], Pobedinsky et al. [2], Kotikov [3], Klubnichkin at al. [4], Pirnazarov et al. [5].

The researches in question of non-rigid properties of soil affected by buildings' and structures' foundations as well as vehicle running gear have found out a variety of mathematical models describing compression and shearing deformation of soil. The rutting process researchers have been mainly conducted from the point of improvement of passability and tow properties of the vehicle and have given no ecological assessment of soil and ground state at logging areas.

The problem of ecological aftermaths of mechanized timber cutting raises a problem of development of theoretical foundations of forecasting of industrial factors' impact on forest soils. At the stage of development of logging area it is necessary to decide upon timber-harvesting equipment and timber-harvesting technology to be used, providing for state of logging area soils not reaching limit values of the biological activity the case when soil damage is not irreversible.

2. Materials and methods

When a vehicle moves, a "core of packed soil" is formed under wheels or tracks in lateral section. This core reminds a wedge and is introduced into a bearing mass shearing away the soil on the sides. That is why general deformation which is a rut depth represents a sum of packing and shearing deformations.

Movement of skidders along skidding tracks is accompanied by accumulation of deformations thus causing deepening of rut depth Bottinelli et al. [6], Cambi et al. [7], Horn et al. [8], Klubnichkin et al. [9]. The number of vehicles' trips along a track, depending on maximum load of running gear, type and state of soil,· leads either to weakening of rut depth growth rate, or vise versa to its progressive growth: in the first case the bearing mass features packing deformations, and in the second case, shearing deformations.

The energy conveyed to soil from a wheel or track is converted into soil deformation energy and is determined by the following formula.

$$A = b \cdot S_{\partial} \int_0^{hk} q \cdot dh \tag{1}$$

A–rolling energy during the cycle time of soil loading by vehicle element running gear;
b–running gear width;
$S_д$–actual trip mileage for a specific period of time;
hk–full deformation (rut depth);
q- pressure.

The analysis of the above equation shows that in lack of slipping when actual trip$S_д$ corresponds to length S_t of surface of contact between running gear and soil, expression (1) describes the energy required for pressing into soil a flat die having a width *b* and length S_t to the depth equal to rut depth.

The final objective of the search is to determine the degree of soil packing under vehicle's impact. Let us take a process of rutting as a consecutive soil deformation by a flat die of the given size and use for mathematical description of that process the provisions of soil mechanics obtained basing on general theory of linear solids which assumes that deformation *h* varies in direct proportion to pressure *q*, that is

$$h = a^*q \tag{2}$$

where: *a* is a coefficient of soil linear deformation dependent on its mechanical properties, size of deformer and other factors.

The soil mechanics applies the following formula to determine coefficient of linear deformation of masses of homogeneous soil

$$\dot{a} = \frac{(1-M^2)}{E_0} \cdot b \cdot \partial \tag{3}$$

where: E_0 - a generic deformation module of quasi-homogeneous semi-space;
M – coefficient of longitudinal expansion of soil;
b – width of loading site;
∂ – coefficient of the form dependent on x= S_t/b.

In this work it is proposed to determine coefficient *a* by the following formula

$$a = 1{,}12\frac{1-M}{E_0} \cdot b \cdot x^{0{,}385} \tag{4}$$

The process of deformation caused by vehicle running gear is non-linear in the severe soil-earth conditions typical for most cutting areas.

To determine general soil deformation, we shall use the following dependence '

$$h = \frac{aq}{1+b'q} \tag{5}$$

hereinafter q is a maximum value of running gear pressure; b' is a value dependent on type of deformation.

In lack of shearing deformations when a rut is formed as a result of soil packing

$$b' = \frac{a}{h_{max}} \tag{6}$$

where: h_{max} is a maximum deformation of soil packing before the state of most dense packing of solid particles at $q \to \infty$.

In case deformation of water-saturated soils, where shearing deformations prevail, the last extend infinitely at pressure q_s representing a limit of bearing capacity. In this case value $b' = -1/q_s$.

Considering that loading of weak forest soils is always accompanied by appearance of packing and shearing deformations, we introduce a formula to calculate generic soil deformation at its one-time loading

$$h = \frac{a \cdot q}{1+\frac{a \cdot q}{h_{max}}} + \frac{a \cdot q^2}{q_s - q} \tag{7}$$

The maximum packing deformation is determined basing on the method of soil equivalent layer proposed by Professor N.A. Tsytovich by the following formula.

$$h_{max} = H_0\left(1 - \frac{P_0}{(1-W)\cdot P_{m6}}\right) \quad (8)$$

where: H_0 – thickness of soil equivalent layer; P_0 – initial density; P_{m6} – soil density in the state of most dense packing of particles; W–soil moisture.

3. The study of accumulation of deformation of soil packing

The complexity of mathematical description of the process of soil deformation at multiple application of load is that as a result of each soil loading its strength properties change and with further vehicle's trips along its tracks they rest on completely different solid mass where the contraction stresses vary and the packing and sharing deformations develop in a different way.

Relatively low elasticity of forest soils makes it possible to consider them as elastic materials where factors of mechanical properties vary with loading application and remain constant after unloading.

Within the frameworks of the indicated assumption there is a possibility to describe accumulation of deformation of soil packing at multiple trips of vehicles along skidding trails by the following expression

$$\sum h_{yn} = \frac{a_0\cdot q\cdot n}{1+\frac{a_0\cdot q\cdot n}{h_{max}}} \quad (9)$$

where: n - number of trips; a_0 coefficient of linear deformation of soil at the beginning of first loading.

Analysis of formula (9) demonstrates that the generic soil packing deformation after n-loading is equivalent to it deformation at one-time application of loading creating pressure $q \cdot n$.

The nature of accumulation of packing deformations is illustrated by the diagram in Fig. 1 which explains the essence of approach to problem solving.

The shearing deformation at similar degrees of loading follows the geometrical progression law

$$\sum \Delta h_C = \frac{a_0\cdot q^2}{q_s - q}\cdot\frac{1-\xi^n}{1-\xi} \quad (10)$$

where: ξ – progression denominator

$$\xi = \frac{2\cdot q_s\cdot q - q^2}{(q_s - q)^2} \quad (11)$$

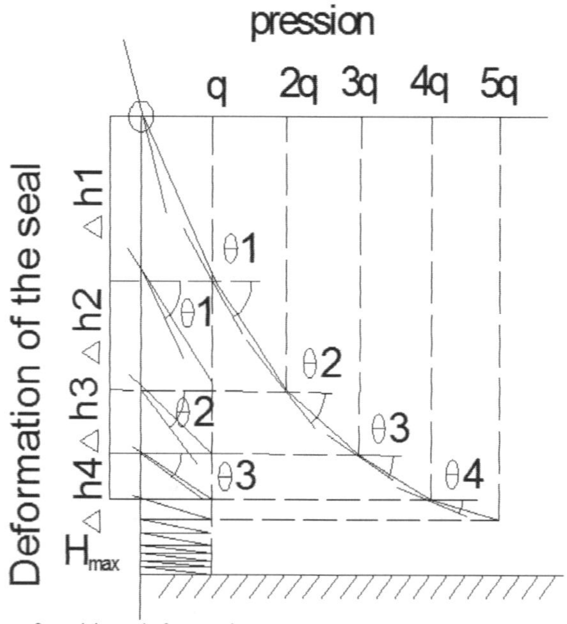

Figure 1. Accumulation of packing deformations

The generic soil deformation after n-trip of a vehicle can be found as a sum of packing and shearing deformations. Then the following expression shall be written down to determine the rut depth

$$\sum \Delta h = \frac{a_0 \cdot q \cdot n}{1+\frac{a_0 \cdot q \cdot n}{h_{max}}} + \frac{a_0 \cdot q^2}{q_s - q} \cdot \frac{1-\xi^n}{\xi} \tag{12}$$

Analysis of equation (12) demonstrates that the nature of rut depth growing at multiple trips of a vehicle along the skidding trail depends on the progression ratio value ξ. At $\xi<1$ the rate of rut depth growing reduces with increase of a number of trips and vice versa, at $\xi> 1$ the rut depth progressively grows.

Since the progression ratio depends on the running gear pressure q and limit of soil bearing capacity, let us review the case when

$$\xi = \frac{2q_s q - q^2}{(q_s - q)^2} = 1 \tag{13}$$

where we find that the rut depth growth rate rises at $q_o= 0{,}293\ q_s$

Relationship $q_o/q_s= K_u$, which characterizes the rate of deformation accumulation, is proposed to be called a criteria of rutting intensity [10, 11].

The average pressure of active-bearing sections of timber harvester tracks is determined by empirical formula

$$q_{cp} = \frac{G}{2b\left(n_k \cdot t + \left(L - n_k t\right) \cdot \sqrt[n_k]{(2q_{co})/q_s}\right)} \tag{14}$$

where: G –full weight of a skidder with a bundle on a butt plate;

n_k – number of support rollers;

L – wheel base;

q_{cp} – average pressure of bearing beam $q_{cp} = G_k / b \cdot L$.

The analysis made in this work demonstrates existing relations between running gear maximum pressure q and average pressure of active-earing sections of tracks q_{cp}, as well as dependence of those relations on the type of soil deformation.

Simultaneous consideration of the impact of a large number of factors which significantly vary along the vehicle path makes us correlate the calculation results with experience with regard to final readings of the rut depth.

4. Conclusion

The analysis made in this work demonstrates existing relations between running gear maximum pressure q and average pressure of active-earing sections of tracks q_{cp}, as well as dependence of those relations on the type of soil deformation.

The rutting process and soil packing degree are significantly dependent on the vehicle's vertical dynamics. Researches demonstrate that the vehicle's speed and design parameters of the running gear affect the values of normal and shear stress formed in the soil.

Implementation of the proposed model shall provide for more reasonable approach to the problem of cutting area development during snowless season, choose rational vehicle systems and technology of their employment.

5. Acknowledgments

Research financed by the Ministry of Education and Science of the Russian Federation project number № 9.8996.2017/8.9.

References

[1] Melekhov I.S. 1980. Forest science. *Forest industry. Moscow. Pages 406.*
[2] Pobedinsky A., Zheldak V. 1989. Features of thinning in forests with limited forest

IOP Conf. Series: Materials Science and Engineering **386** (2018) 012025 doi:10.1088/1757-899X/386/1/012025

management. *Forestry. Vol. 9. 24-27.*
[3] Kotikov V.M. 1995. Impact harvesting machines on forest soils. *Dissertation of the doctor of technical sciences. Moscow state forest university, Pages 214.*
[4] Klubnichkin E., Klubnichkin V., Krylov V., Kondratyuk D. 2012 To justify the specific pressure of a tracked forestry tractor. *Forestry Bulletin. Vol. 8. 48-50.*
[5] Pirnazarov A., Sellgren U. 2015. Reduced testing and modelling of the bearing capacity of rooted soil for wheeled forestry machines. *Journal of Terramechanics. Vol. 60, 23-31.*
[6] Bottinelli N., Hallaire V., Goutal N., Bonnaud P., Ranger J. 2014. Impact of heavy traffic on soil macroporosity of two silty forest soils: Initial effect and short-term recovery. *Geoderma. Vol. 217–218, 10-17.*
[7] Cambi M., Certini G., Neri F., Marchi E. 2015. The impact of heavy traffic on forest soils. *Forest Ecology and Management. Vol. 338, 124-138.*
[8] Horn R., Vossbrink J., Peth S., Becker S. 2007. Impact of modern forest vehicles on soil physical properties. *Forest Ecology and Management. Vol. 248, Issues 1–2, 56-63.*
[9] Klubnichkin E., Klubnichkin V. 2017 Theoretical foundations of preliminary assessment of the impact of the timber harvesting vehicles on forest soils in the course of logging. *Proceedings of the 19th International & 14th European-African Regional Conference of the ISTVS*
[10] Klubnichkin V., Klubnichkin E., Kotiev G. 2017 Developing a method of theoretical evaluation of the tracked timber harvesting machine undercarriage element loading. *Proceedings of the 19th International & 14th European-African Regional Conference of the ISTVS*
[11] Klubnichkin E., Klubnichkin V., Nakaznova O., Naumov V., Kotiev G., Belyakov V., Zezyulin D., 2016. Influence of the distribution of normal pressures of caterpillar forestry machines LZ-4 and LZ-5 for track formation. *Logging journal. Vol. 6. Issue 3, 167-176.*

IOP Conf. Series: Materials Science and Engineering **386** (2018) 012026 doi:10.1088/1757-899X/386/1/012026

The impact of changing the type of understeer on vehicle handling

Yu I Molev[1], M Cherevastov[2], A V Sogin[1], A S Slyusarev[3], S S Ivanov[1]

[1] Nizhny Novgorod State Technical University n.a. R.E. Alekseev, professor, Russia
[2] Nizhny Novgorod, Russia, Nizhny Novgorod State Pedagogical University n.a. K. Minin, graduate student, Russia
[3] Nizhny Novgorod, Russia, Volga State University of water transport

E-mail: moleff@yandex.ru[1], chermaxim_1978@mail.ru[2]

Abstract. The transient response of a truck with a step control action on the steering studies in the paper. The car is represented by a one-mass flat design model. The equations of the machine motion are written in the form of stability derivatives. Calculation of direct estimates of the quality of the transition process on the angular velocity of the car at a stepwise (step) control action is made. The influence of the of the rotation type change on the car handling is presented.

1. Introduction

Studying the car handling, it can be noted that a number of authors indicate that when designing a car it is necessary to ensure its understeer throughout the range of mass from curb to maximum allowed. However, often, in the operation of trucks those cases are allowed where the cargo weight exceeding the established capacity. At the same time, it is possible to change the rotation type from insufficient to excessive. In this paper, the case of such a transition is considered. In this case, the vehicle handling will be evaluated on the basis of direct assessments of the quality of the vehicle transient response for different initial traffic conditions.

In this paper, we take the linear one-mass plane model of the car as the basis (the motion equations are presented in the form of stability derivatives [1]). Let us recall briefly the main assumptions typical for this calculation model. The drag coefficient of the side wheel drive is a constant. There is no redistribution of vertical loads on the wheels of one axis, and the position of the car gravity center when it moves is not changed. The angles of the wheels slip and the steering angles of the steered wheels have small values. Applying the classical traditional approach of the automatic and control regulation theory, the car transient response is determined by an instant change in the angle of rotation of the driven wheels. When calculating, we also take the following: the translational speed of the car remains constant for each movement case, only disturbances from control are taken into account, as for aerodynamic disturbances and road effects, they are not taken into account. On the basis of direct assessments of the quality of the transition process, under different conditions, theoretical conclusions about the impact on the quality of the transition reaction of the car on the angular speed of rotation changes in the type of rotation will be made.

2. Main part

In this article the following notations are accepted:

a, b – distance (horizontal) from the centre of gravity of the vehicle to the front (rear) axle [m];
θ – angle of rotation of operated wheels [rad];
ω – angular velocity of the car rotation [1/s];
s – Laplace\` operator;
C_1, C_2 – coefficient of resistance to the tire of the front (rear) wheels [N/rad];

Jz – moment of inertia of the vehicle, relative to the vertical axis passing through its center of gravity [kg·m^2];
M – vehicle weight [kg];
V – forward speed of the car [m/s];
$Y_\beta = C_1 + C_2$- derivative of the resistance (for lateral force), defined by the presence of the slip of the car wheels [N/rad];
$Y_\omega = \frac{1}{V}(aC_1 - bC_2)$- derivative of the resistance (for lateral force), defined by the rotation of the vehicle about a vertical axis passing through its centre of gravity [N·s];
$Y_\theta = -C_1$- the derivative of stability (for lateral force), determined by the rotation of the driven wheels [N/rad];
$N_\beta = aC_1 - bC_2$- derivative of stability (by turning moment), caused by the presence of the slip of the car wheels [(N·m)/rad];
$N_\omega = \frac{1}{V}(a^2C_1 + b^2C_2)$- derivative of stability (by turning moment), caused by the rotation of the car, with respect to the vertical axis passing through its center of gravity [N·m·s];
$N_\theta = -aC_1$- the derivative of stability (by turning moment), caused by the rotation of the driven wheels [(N·m)/rad].

For studying, we considered a two-axle truck with a rear driving axle, which has the technical characteristics specified in table 1.

Table 1. General technical characteristics of the truck.

Wheelbase	2,9 m
The coefficient of cornering resistance	$-40\left[\frac{kN}{rad}\right]$
Gross vehicle weight / Curb weight	3500 kg / 1850 kg
Curb weight distribution (front axle [M_1] / rear axle [M_2])	1050 kg / 800 kg
Total mass distribution (front axle [M_1] / rear axle [M_2])	1200 kg / 2300 kg

Having the value of the coefficient of resistance to the lateral lead of one tire, taking into account that all the wheels are equipped with the same tires with the same parameters, we obtain for the axes:

$$C_1 = 2C = -80\left[\frac{kN}{rad}\right];\ C_2 = 4C = -160\left[\frac{kN}{rad}\right]$$

The necessary values for the position of the centre of gravity of the vehicle at different values of its mass and moment of inertia are presented in table 2.

Table 2. Individual parameter values.

Vehicle weight (*M*), kg	Position of the car centre of gravity		Vehicle moment of iner (J_z), kg m^2
	a, м	*b*, м	
1850	1,256	1,644	3820
3500	1,91	0,99	6620
3700	1,96	0,94	6820

Let's make a characteristic equation of a truck and not take its roots. Using the work of [2] we have:

$$s^2 + B_1 s + B_2 = 0 \tag{1}$$

where $B_1 = -\left(\frac{N_\omega}{J_z} + \frac{Y_\beta}{MV}\right)$; $B_2 = \frac{N_\beta}{J_z} - \frac{N_\beta Y_\omega}{MVJ_z} + \frac{Y_\beta N_\omega}{MVJ_z}$.

Consider the case when the roots of characteristic equation (1) are real, then the equality is secured:

$$\omega(t) = E_1 + E_2 e^{s_1 t} + E_3 e^{s_2 t} \tag{2}$$

where $A_1 = \frac{N_\theta \theta_{yc}}{J_z}$; $A_2 = \frac{\theta_{yc}\left(Y_\theta N_\beta - N_\theta Y_\beta\right)}{MVJ_z}$;

$s_{1,2} = \frac{-B_1 \pm \sqrt{B_1^2 - 4B_2}}{2}$;

$E_1 = \frac{A_2}{s_1 s_2}$; $E_2 = \frac{-\left(E_1 s_2 + A_1\right)}{s_2 - s_1}$; $E_3 = -\left(E_1 + E_2\right)$.

For this case the quality of the transition process is evaluated using only control time and by employing the traditional approach of the automatic control and regulation theory we will define the time interval from the moment of application of the impact to the moment when the value becomes less than the value of 0.1 ω_{us} at all subsequent moments of time. The analysis of the expression (2) shows that if the roots of the characteristic equation of the truck are valid and negative, then with increase in time the value of the angular velocity of rotation of the machine will tend to the value of E_1 which equals to the magnitude of the steady-state angular velocity (ω_{us}) after the completion of the transition process.

Now let us consider the situation when the roots of equation (1) are imaginary. In this case, the transient response of the vehicle is oscillatory and is determined by:

$$\omega(t) = T_1 + Ze^{\frac{-B_1}{2}t}\sin(\omega_k t) + T_2 e^{\frac{-B_1}{2}t}\cos(\omega_k t) \tag{3}$$

where $\omega_k^2 = B_2 - \left(\frac{B_1}{2}\right)^2$; $T_1 = \frac{A_2}{B_2}$; $T_2 = -T_1$; $T_3 = A_1 - B_1 T_1$; $Z = -\frac{T_2}{\omega_k}\left(\frac{B_1}{2} - \frac{T_3}{T_2}\right)$.

Here we evaluate the quality of the transition process by the previously mentioned control time and the magnitude of overshoot (σ) which is numerically equal to the ratio of the difference of the maximum value of the angular velocity of rotation of the vehicle rotation during the period of the transition process and the steady-state value of angular speed of after the end of the transition process to the steady-state value of angular speed.

Using expressions (2) or (3), for the corresponding roots of the characteristic equation of the truck, using the necessary data from tables 1 and 2, we counted the values of stability derivatives and constructed the car transient responses to the instantaneous change in the angle of rotation of the driven wheels from 0 to 0.17 radians at different values of the truck mass.

Figure 1 shows the transient response for curb weight at three different translational speeds of the vehicle.

IOP Conf. Series: Materials Science and Engineering 386 (2018) 012026 doi:10.1088/1757-899X/386/1/012026

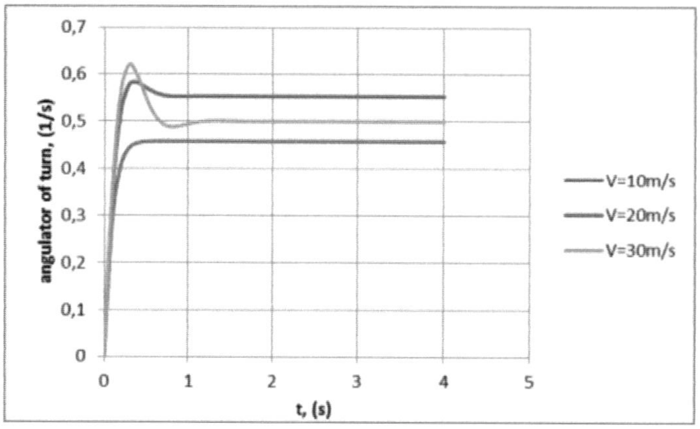

Figure 1. The nature of the transition process.

In figures 2 and 3 the dependence of the regulation time and over-regulation on different values of the translational speed is presented.

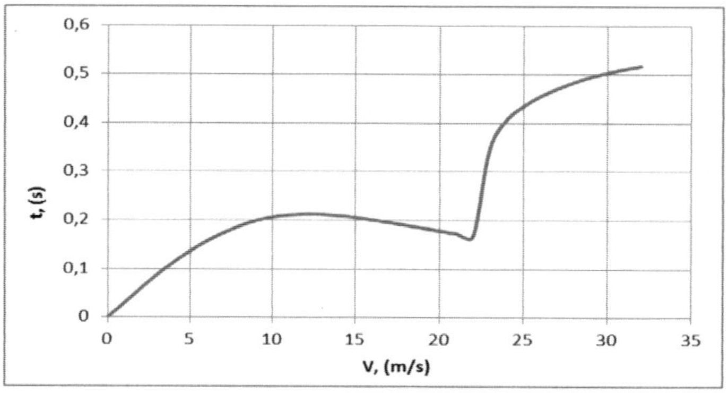

Figure 2. Control time.

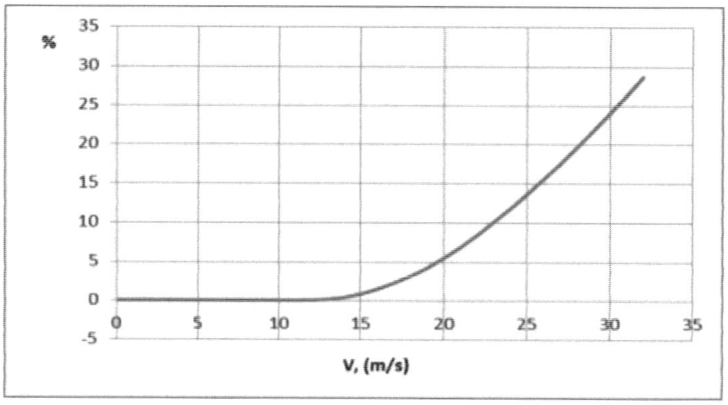

Figure 3. Overshoot.

In the same sequence in Fig. 4-Fig. 6 the corresponding dependences for the maximum permissible mass of the vehicle are given.

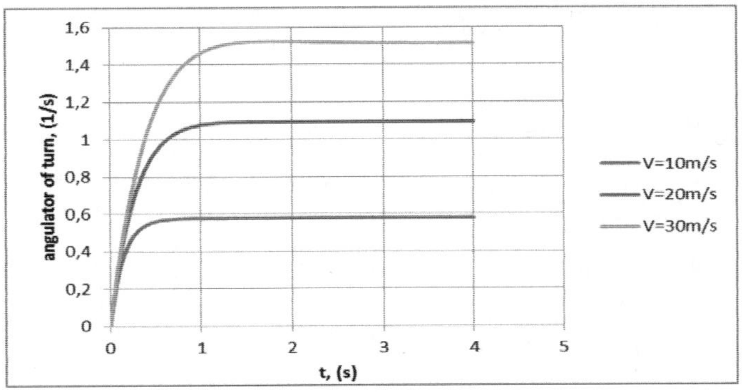

Figure 4. The nature of the transition process.

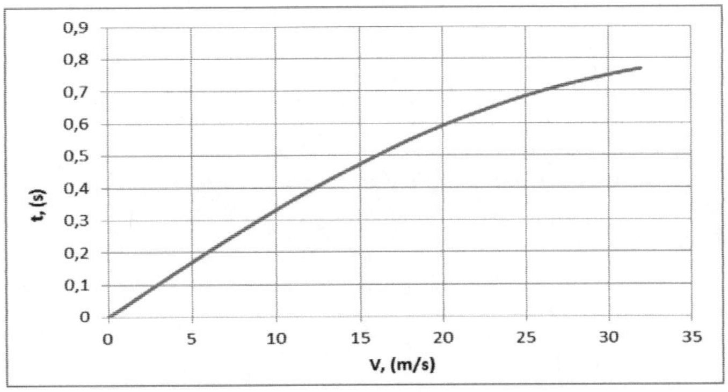

Figure 5. Control time.

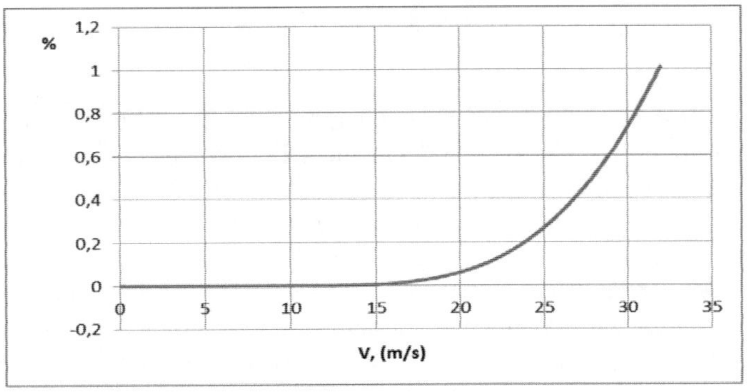

Figure 6. Overshoot.

Next, we consider the case for mass of 3700 kg. Here the character of the transition reaction for different speeds will have the following form shown in Fig. 7.

IOP Conf. Series: Materials Science and Engineering **386** (2018) 012026 doi:10.1088/1757-899X/386/1/012026

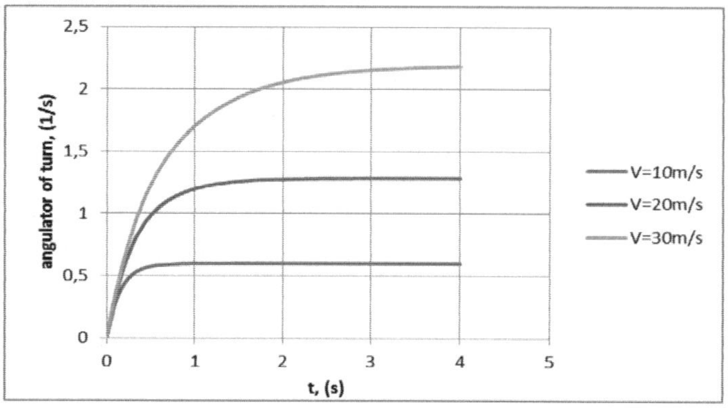

Figure 7. The nature of the transition process.

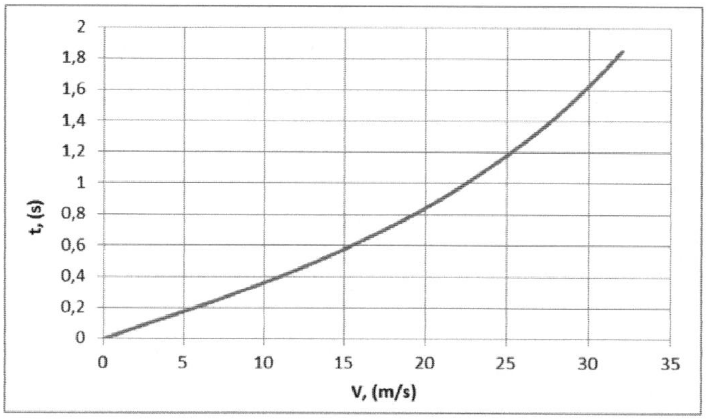

Figure 8. Control time.

3. Result of work.

The analysis of the relationships of characteristic quantities and for different masses of the car shows that within the permissible mass [1850…3500 kg] there is understeer, in the case of exceeding the maximum allowed masses we have oversteer. The conducted direct evaluations of the quality of the transient response of the angular velocity of the car showed that at curb weight the control time determining the speed of the linear dynamic system (truck) is in the range from 0 to 0.52 seconds for speeds from 0 to 32 m/s, and the value of overshoot characterizing the oscillation of the system reaches 29% at a speed of 32 m/s at the maximum permitted mass, the vibration of the transient reaction is weakly expressed, and the maximum control time increases to 0.76 seconds. At the onset of oversteer, the law of change of the transient reaction is monotonic aperiodic in nature, with the system speed significantly reduced, and the control time begins to increase to the value of 1.86 seconds.

In conclusion, it should be noted that the change in the type of rotation from insufficient to excessive leads to decreasing in the operating speed of the truck as a linear dynamic (mechanical) system and inevitably has a negative impact on its handling.

4. References.

[1] Milliken U. F. and Uitkom D. U. Application of the General theory of stability and controllability of cars to their design / Controllability and stability of the car. Collection of articles / edited by A. S. Litvinov. M: MASHGIZ, 1963. - P. 145-204.

[2] Molev, Y. I., Cherevastov M. G. Theoretical calculation of the transient response of the vehicle at the given function perturbations [Text] / Organization and traffic safety: materials of XI international scientific-practical conference (March 15, 2018): in 2 vol / ed. edited by A. D. Zakharov. – Tyumen: TIU, 2018. / Vol. 2. – P. 89 – 95.

IOP Conf. Series: Materials Science and Engineering **386** (2018) 012027 doi:10.1088/1757-899X/386/1/012027

Virtual Reality and Navigation subsystems of the Interactive System for Road Safety Improving

M E Eliseev T N Tomchinskaya E D Galkina N Yu Dudakov, R A Musarsky

Nizhny Novgorod State Technical University n.a. R. E. Alekseev, Russia

E-mail: eliseevmic@mail.ru

Abstract. The article considers the Interactive System for Road Safety Improving of a large city. Its virtual reality and navigation subsystems, their interaction with each other and with the main system are described in detail. The navigation subsystem informs drivers about potentially dangerous sections of the road in real time. The virtual reality subsystem allows you to visualize various mechanisms of emergency situations, connect them with the landscape, various conditions of the urban environment, and a road accident statistics map. The alpha version of the software is being tested in Nizhny Novgorod, Russia. It is planned to use the system by all subjects of the road process: drivers, pedestrians, enterprises involved in transportation, road services, road traffic organizers.

1. Introduction

The problem of accident rate has become especially dangerous. According to statistics, about 1,2 million people die every year in road accidents (3300 people a day). From 20 to 50 million are not fatal injuries. At the same time, over three quarters of all road accidents are related to violations of the Road Traffic Rules by drivers of vehicles. Drivers with a driving experience of up to three years make up a large percentage of participants in road accidents. Virtual Reality and Navigation subsystems of the Interactive System for Road Safety Improving of a large city (further ISIRS) are primarily aimed at reducing accidents in this category of drivers. ISIRS is a system for collecting, storing and processing information related to road accidents, as well as for feedback to road users.

Similar systems are being developed by a number of authors [1-14]. For example, Hanshin Expressway Company Limited considers the risk of an accident as one of the important concepts of its safe driving assistance and traffic management for the next generation. The company analyzes its various data on the driving conditions it has as an operator network and road accident records so as to be able to predict traffic accidents for these driving conditions. The company suggests using information about the risks of accidents for comfortable and safe driving [15].

Interesting studies offer an architecture for interactive motion-based motion modeling Environment. In order to enhance modeling realism involving actual human beings, the proposed architecture integrates multiple types of simulation, including: motion-based driving simulation, pedestrian simulation, motorcycling and bicycling simulation, and traffic flow simulation. The architecture has been designed to enable the simulation of the entire network; as a result, the actual driver, pedestrian, and bike rider can navigate anywhere in the system. In addition, the background traffic interacts with the actual human beings [16].

The principal differences of the proposed system are as follows: 1) the most complete collection of information related to accidents; 2) structuring information; 3) autonomy; 4) adaptability; 5) statistical validity of the obtained results of information processing; 6) transfer of information to the participants of the road process in a form convenient for make use of [17].

Autonomy here means that once configured for a particular city system does not require regular support and can operate in a fully automated mode. Naturally, the system is updated with new information and its work can be adjusted by specialists. By adaptability we mean that the ISIRS can be configured for any large city by changing a number of parameters. Under the form convenient we mean that it may be used for trainees in driving schools, there is information videos linked to specific sections of the road infrastructure, for drivers there is navigation application hints, for traffic experts - accident monitoring datarepresented in information banners on the map, charts, graphs and tabular forms.Figure 1 shows one example of the presentation of information about foci of road accidents.

Figure 1. Information window of the detailed information of a node

The following hierarchy is assumed (Figure 2):
- the main urban information management system "Smart City";
- transport information subsystem;
- ISIRS;
- subsystems of virtual reality and navigation.

Information from sensors of various types and from various city services enters the information-management system "Smart City".That part of it that is used in transport processes is delivered to the transport subsystem, if required, in a revised form.In the transport subsystem, the data are analyzed and their part that is related to road safety is transmitted to the ISIRS.

It processes the information received and transmits it in the desired form to the subjects of the road process: drivers, pedestrians, enterprises involved in transportation, road services, road traffic organizers directly, or returning to the transport subsystem, but in a different form. In its turn, the transport subsystem can either use the data internally, or return it to the global urban information management system "Smart City", which will forward the information to the necessary subjects of urban control processes. Such a delimitation of the functions of the systems will allow the most complete and accurate processing of data entering the global system.

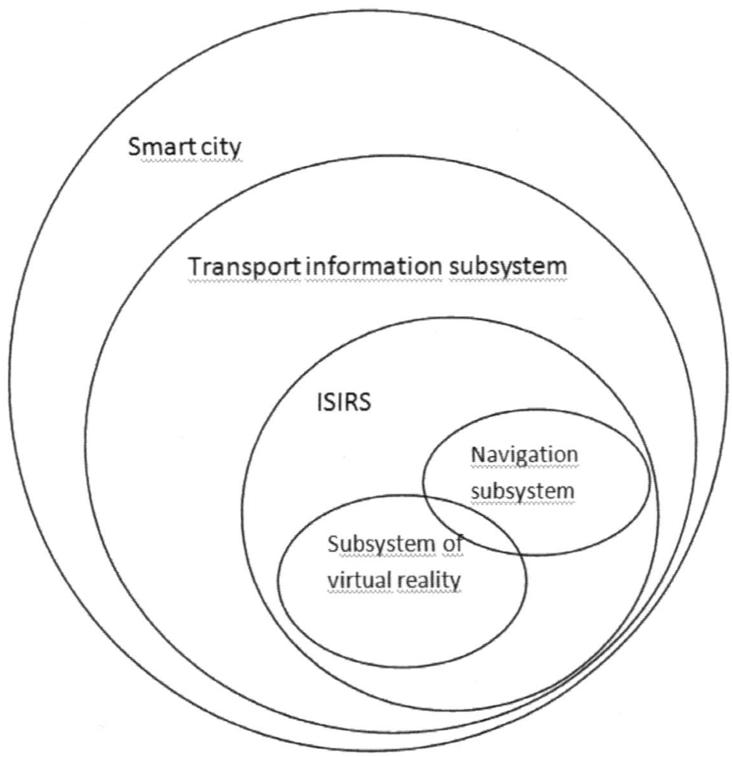

Figure 2. Hierarchy of urban information management systems

Let us confine ourselves to one small illustration, to explain the nature of the interaction of systems.Let the weather forecast for the next day come from the meteorological stations in the city.Since this information affects transport processes, it is transferred to the transport subsystem.In view of the fact that the weather data is related to road safety, they are transmitted to the ISIRS, where they loads into the subsystem of weather data analysis.If it is revealed that these data correspond toone of the criteria of a dangerous combination of weather factors, the following actions will be performed:

1. In the ISIRS annotation subsystem information messages will be created. They will be transferred to the ISIRS subsystems, which will provide feedback to the subjects of the road process, in particular, the navigation subsystem. The navigation subsystem will add this information to the information messages for drivers when passing the accident foci related to the appropriate combination of weather factors.

2.The navigation subsystem will transfer the data in a revised form to the transport subsystem that will send the information to the appropriate subsystems, for example, the passenger transport subsystem, and then send them to the "Smart City". If this information requires the connection of city services (for example, a heavy snowfall is predicted), data, in a revised form will come to the subjects of urban processes.So, in the case of forecast data, "heavy snowfall", road services associated with snow removal will be put in a ready state.A very simplified example is given, as in ISIRS, of course, it is not simply a "snowfall" being considered, but a set of forecast data (air temperature, pressure, etc.) verifies the fulfillment of one of the criteria.

General view of the criteria: $k_1, k_2, \ldots k_6$ – values of the variables of the weather factor (k_1 - predicted temperature, k_2 - predictedpressureetc.). Factors are previously selected using factor analysis, $\rho_1, \rho_2, \ldots \rho_6$- values of correlation coefficients between the relevant weather factor and the daily number of accidents. For a more accurate prediction, the entire temperature range was divided into several intervals. For each interval the criterion is:

$$k_1 \rho_1 + k_2 \rho_2 + \ldots + k_6 \rho_6 < K_0 \tag{1}$$

where K_0 – is the threshold value of the criterion, which is selected experimentally for each temperature range.

In order to understand the internal interaction of components in ISIRS, we list its main subsystems: topographic analysis, route construction, accident monitoring, 3D modelling, weather data analysis, annotation, client-server application, navigation, virtual reality.

In this paper, the last two subsystems of this list are presented.

2. The navigation subsystem of ISIRS

For feedback from drivers, as one of the main participants of the road process, information on accidentally-dangerous road factors should be transmitted to them in a form convenient for perception.One of the most suitable means of transmitting these data, along with information monitors, are navigators and navigation applications.

We will discuss in detail how the interaction of various subsystems of the ISIRS takes place to form navigational hints.

1. The driver indicates the start and end points in the navigation application.

2. This data, together with information about driving experience, gender and age of the driver (if authorized) are transferred to the subsystem of the client-server application that refers to the subsystem of the construction of routes.At this stage, the path is built in the "json" format.A fragment of the encoding of this route in the"json" format is shown in Figure 4.

[{"lat":56.34777875514101,"lng":43.86947958940115,"street":"\u041a\u041e\u041c\u0418\u041d\u0422\u0415\u0420\u041d\u0410","house":"105","full_count":179,"distance":6.8929747339046325e-6,"description":"\u0412\u0441\u0435\u0433\u043e \u0437\u0430 \u0432\u044b\u0431\u0440\u0430\u043d\u043d\u044b\u0439 \u043f\u0435\u0440\u0438\u043e\u0434 \u0432\u043e\u0447\u0430\u0433\u0435

Figure 3. Fragment of "json" file, encodedthe selected path

3. Routes in the form of a set of points and polylines are transferred to the topographic analysis subsystem, in which the places of concentration of accidents (furtherthe accidents foci) on the route are determined.

4. There is an appeal to the subsystems for monitoring the accident rate and analyzing the weather data. At this stage, information is requested about accidents foci along the way.

5. Information about the foci, weather data and accident monitoring data, together with the driver's personal information (experience, gender and age), is transmitted into the annotation subsystem.

The formation of an annotated message for a specific focus is performed using the following algorithm.

2.1. Algorithm for generating an annotated message

1. If there is an expert comment, it will be given the highest priority, and it will be automatically added to the message.

2. The weather data are analyzed. According to the table of criteria, it is checked whether the existing combination of weather factors satisfies at least one of the criteria of the form (1), if yes, then this

IOP Conf. Series: Materials Science and Engineering **386** (2018) 012027 doi:10.1088/1757-899X/386/1/012027

criterion will be included in the processing, if not, then the weather data will not be taken into the annotation.

3. Road infrastructure data tables are checked: road works, idle traffic lights, etc., if there is a focus data in them, the corresponding information is added to the message.

4. The data of the monitoring subsystem are checked, if a sharp growth of accidents of certain categories is noticed in this center, then at the annotation this data will have the next priority level.

5. If during the last week and the month sharp increase in the number of accidents is not observed, the annotation will use data for the last 3 years: the accident factors and categories of drivers for which the average accident rate is significantly exceeded and a message will be built on them.

Table 1. Table of possible values of accident factors.

№	Data	Value
1	Expert's comment	text
2	Weather criteria 1	0 or 1
3	Weather criteria 2	0 or 1
4	Weather criteria 3	0 or 1
5	Weathercriteria 4	0 or 1
6	Weathercriteria 5	0 or 1
7	Roadinfrastructure	0-4 (the digit corresponds to the factor: 0 - factors are absent, 1 - road works, 2 - not working traffic light and etc.)
8	Monitoringdatafactor	0-9 (the digit corresponds to the factor: 0 - factors are absent, 1 - distance, 2 - priority and etc.)
9	Monitoringdatadrivercategory	0-10 (0 - not allocated categories, 1 - male, 2 - female, 3 - driving experience 0 full years and etc.)
10	Generaldatafactor	0-9 (the digit corresponds to the factor: 0 - factors are absent, 1 - distance, 2 - priority and etc.)
11	Generaldatadrivercategory	0-10 (0 - not allocated categories, 1 - male, 2 - female, 3 - driving experience 0 full years and etc.)

Let us dwell in more detail on the technique of calculations. The marker of the cause of the accident rate was calculated. This is a row matrix, the elements of which are the portions:

$$K_i = \frac{n_{io} \times N}{N_o \times n_i} \tag{2}$$

where$N$ – total number of RA, n_i – the number of RA with the *i* factor, N_o – total number of RA in the node, n_{io} – the number of RA in the node with the *i* factor. List of factors-conditions associated with RA or types of RA, is set in advance. When annotating the largest coefficients K_i in groups: driving experience, gender, driver age, concomitant factor, weather factor are taken into account. In this case, weather factors are displayed only when the conditions of clause 2 are fulfilled, other factors only if their value is not less than the specified threshold, the latter is proposed to be taken equal to 1.2, i.e. the average accident rate is exceeded by 20%.

As a result, annotations are formed for each focus along the route. The message created for the focus is shown in Figure 3.

IOP Conf. Series: Materials Science and Engineering **386** (2018) 012027 doi:10.1088/1757-899X/386/1/012027

1. {lat: 56.35200119018555, lng: 43.86750030517578, street: "КОМИНТЕРНА", home: "172", full_count: 18244,...}
2. description:"При проезде очага будьте предельно ВНИМАТЕЛЬНЫ!!!При образовании затора обратите внимание на боковой интервал. Здесь многие аварии происходят из-за нежелания уступить другому."
3. dtp_member:[{rate: "2.3577", color: "red", experience: "менее 1 года", age: "41 - 60 лет", gender: "female",...}]
4. full_count:18244
5. home:"172"
6. lat:56.35200119018555
7. lng:43.86750030517578
8. node_count:53
9. sop_factor:[{rate: "0.2303", color: "gray", name: "Сухое", cnt_full: 49335, cnt_node: 33}]
10. street:"КОМИНТЕРНА"
11. type:[{rate: "1.8490", color: "red", name: "Боковойинтервал", cnt_full: 2234, cnt_node: 12}]
12. weather:[{rate: "0.2133", color: "gray", name: "Пасмурно", cnt_full: 32281, cnt_node: 20}]

Figure 4. Server Response Example - Information about the accident focus

6. Route data including annotations for foci are sent to the navigation application.
7. When approaching a crossroad or a linear section of a route with an annotated message, the standard audio message is supplemented with the corresponding information. Intheexampleabove, itlookslikethis:

"When passing through the focus, be extremely ATTENTIVE! When forming a traffic stack, pay attention to the lateral interval. Here, many accidents occur because of reluctance to give way to another"

Note that without authorization (driving experience, gender and age), the message would look different:

When forming a traffic stack, pay attention to the lateral interval. Here, many accidents occur because of reluctance to give way to another"

This is due to the fact that in the topographic analysis subsystem it is revealed that for this category of drivers this focus has exceeded the threshold value of the ratio of focal and general data.

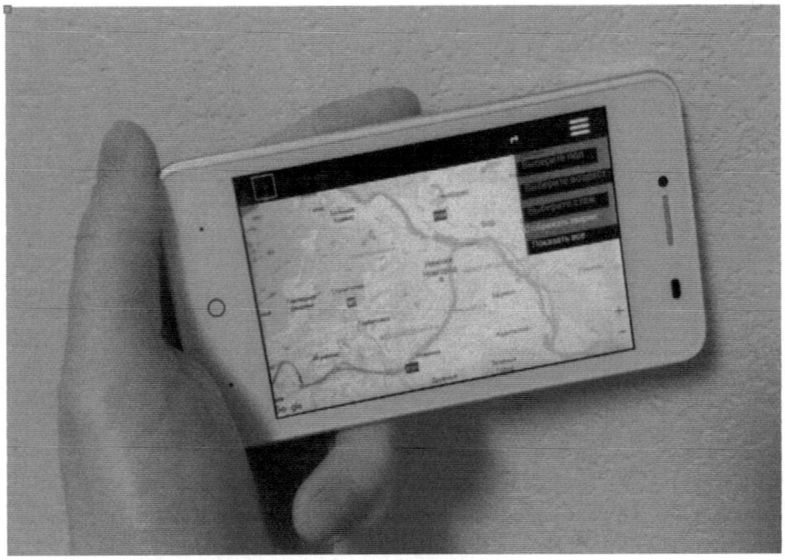

Figure 5. Abuildingroutemenu

The navigation subsystem is under development. The menu for building a route for a mobile application is shown in Figure 4.

3. The subsystem of virtual reality

In this subsystem, two ways of ensuring road safety are considered.

1. Preview an unfamiliar section of the path using augmented reality. Pointing the camera smartphone on a map of the city, for example, Yandex map, the driver can see a realistic model of a specific section of the road infrastructure on the smartphone screen with all the signs and markings.

2. By immersing himself in virtual reality, the driver can see an accident that occurred on a particular section of a realistically modeled part of a particular cityin the stereo mode. You can see this situation from different points of view, like the victim, the culprit or the casual witness and understand all the reasons for this accident.

Based on the information received during the work of the topographic analysis subsystem, a list of the most accident-prone sections of the road infrastructure (foci of accidents) is singled out. For each of the foci, in accordance with the formula (2), the factors that will be the parameters of the simulation are identified.

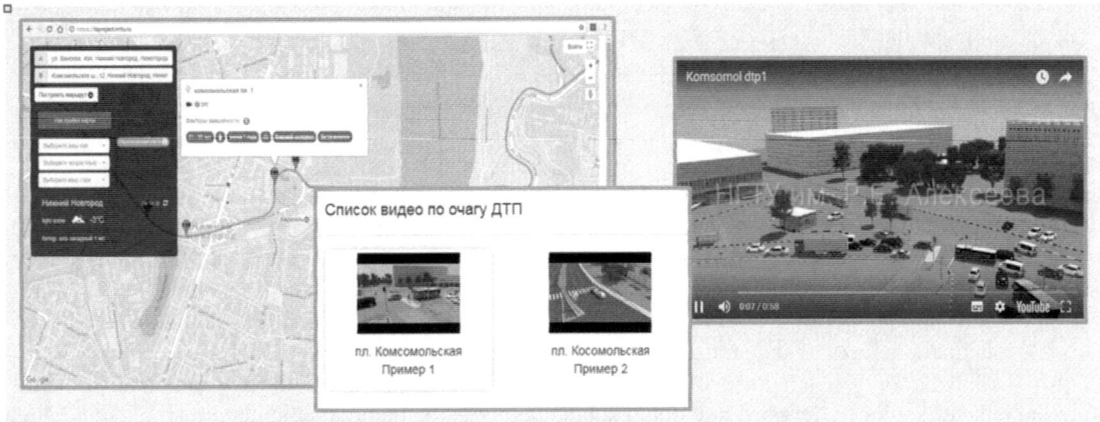

Figure 6. Operation of the ISIRS topographic analysis subsystem

IOP Conf. Series: Materials Science and Engineering **386** (2018) 012027 doi:10.1088/1757-899X/386/1/012027

1.1. The most dangerous parts of the city are identified, corresponding to each factor, and dangerous road situations are modeled on these sections. Recordedvideosshowing:
1) erroneous behavior of the driver and
2) how to act properly in a similar situation. A video bank has been prepared, which stores videos demonstrating traffic rules. Figure 5 shows the accident-prone areas with a factor - "Pedestrian accident".

3.1. Augmented Reality Block
To implement the application on the Android platform, the integrated development environment of Unity 3D is used. Scenes with traffic accidents in the most dangerous parts of the city, previously developed in the technology of low-poly modeling in the tool environment of AutodeskMaya and 3Ds Max, and then converted to a Unity 3D environment.

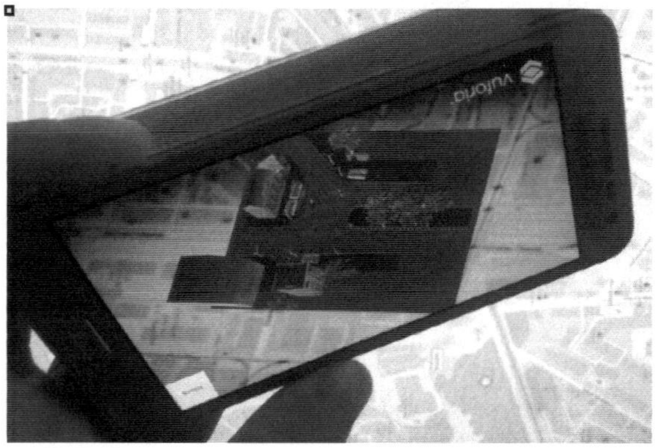

Figure 7. Uploading a scene to the smartphone by the map label

In the application, several blocks are created, each block is responsible for a specific section. An integrated method of creating a program is used, standard modules are used, and the links between them are written using the C # programming language. In the AR, augmented reality, the Vuforia plug-in launches a mechanism for tags searching, created in advance on the basis of maps and linking a model or a tooltip to it. The scene with the most dangerous areas is integrated into the map. The Vuforia plug-in recognizes the image of one of the marks on the map received from the smartphone's camera, based on information about the position of the marker in space.

Then, the calculation of the reference points with simultaneous tracking is performed, thereby the three-dimensional object is oriented in space.

The application projects the corresponding model onto the label, and the scene of the accident case is loaded on the screen. The scene is played in the third person view mode (Figure 6).

3.2. VirtualRealityBlock
Block VR (virtual reality) is designed to view the model in virtual reality mode without using a label. From the list of suggested scenes, the user selects the model of interest. The screen of the smartphone or tablet is loaded with a scene, with the ability to view it from different angles and from different road users with full or partial immersion.

For a full immersion, a virtual reality helmet is required. A virtual camera is added to the scene. The choice is made in favor of the product Fibrum SDK, because it has all the necessary settings, and it does not overload the system with additional plug-ins. Using the Fibrum SDK object, two cameras are installed, for each eye, respectively, the distance between the cameras, the distortion of the image and

IOP Conf. Series: Materials Science and Engineering **386** (2018) 012027 doi:10.1088/1757-899X/386/1/012027

the required cylindrical distortion. To manage the character standard internal plug-in Unity 3D – Characters Controller is used.

When you download a scene on a smartphone in the virtual reality mode, you can view the situation from various faces of road users, whether the culprit, the victim or the free movement of the scene. When the user is in the free movement mode, he can look around, explore the terrain and the road infrastructure surrounding it, or navigate around the stage using the standard control joystick (Figure 7). Buttons for selecting the view mode are at the top of the screen. Back button - to exit the scene, to the main menu (Figure 8).

Figure 8. Navigation and orientation control buttons in the space

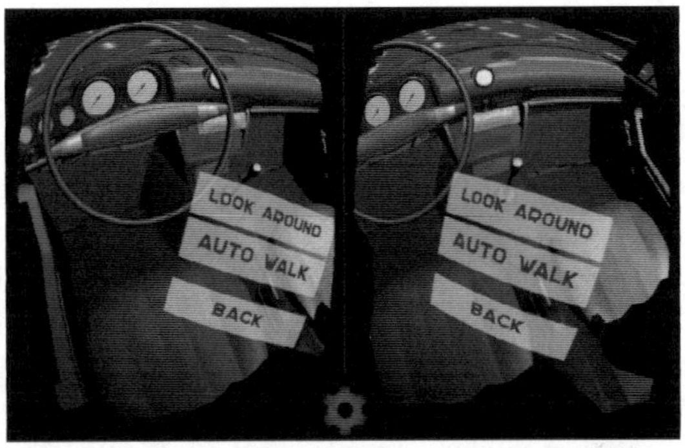

Figure 9. Operation of two cameras for stereo imaging

To switch to viewing a scene from another participant of the road, the user needs to "look" at the corresponding button on the screen. For a more complete immersion, objects inside the scene cast a shadow from static and dynamic light sources. To do this, a 3D object cylinder is added to the parent prefect of the Rigid Body FPS Controller. It casts a shadow from the character, but in order for the polygons of the object to fall into the camera, it is excluded from the renderer. To make the body have its volume and collide with other objects, a standard MeshCollider script is added, which makes all faces of the character physical and does not allow it to pass through the walls. The same plugin should be on objects, through which the user is forbidden to pass.

4. Conclusion

A pilot project of an interactive system for improving road safety, a navigation and a virtual subsystem for informing drivers, is presented. The authors propose to use information on the risks of road accidents and for the realization of safe and comfortable traffic. The ways of practical information service for drivers to choose a route and correct behavior in emergency situations are offered.

In the paper, the application of virtual and augmented reality technologies during driver training in real emergency areas of the city is investigated. To build scenarios for the creation of accident-dangerous cases, statistics and situational descriptions of accidents is used. The mode of dynamic training in the mode of virtual and augmented reality with various degrees of immersion is offered.

We consider the technology for implementing an application for a mobile platform that can implement driver training scenarios in emergency areas of the road infrastructure, in real-life situations that are difficult to drive in several virtual and augmented reality modes. The implementation is performed in the Unity 3D integrated development environment. Testing was conducted in three driving schools of the city. The greatest interest was shown by driving schools (93%) and drivers with a driving experience of up to three years (82%).

The navigation subsystem, implemented as a mobile application, will allow drivers to receive timely information about emergency-dangerous factors in a form convenient for perception, which, according to the authors, will help them to feel more confident on the road and reduce the risk of road accidents.

References

[1] De Lucia B H and Scopatz R A E-Crash: The Model Electronic Crash Data Collection System. *Washington: National Highway Traffic Safety Administration*. 2010

[2] Espie S, Gattuso D, Galante F. A Hybrid Traffic Model Coupling Macro and Behavioral Micro Simulation *Proceedings of the 85th Annual Meeting Transportation Research Board* USA Washington 2006

[3] Graettinger A, Lindly J K and Mistry G J Display and Analysis of Crash Data *University Transportation Center for Alabama*Tuscaloosa AL 2005

[4] Graettinger A J, McFadden J and Rushing T W Evaluation of Inexpensive Global Positioning System Units to Improve Crash Location Data *Transportation Research Record: Journal of the Transportation Research Board* (1746) 2001 pp 94–101

[5] Harkey D L Evaluation of Truck Crashes Using a GIS-Based Crash Referencing and Analysis System *Transportation Research Record: Journal of the Transportation Research Board* (1686) 1999 pp 13–21

[6] Khan G, Santiago-Chaparro K R, Qin X and Noyce D A Application and Integration of Lattice Data Analysis Network KFunctions and Geographic Information System Software to Study Ice-Related Crashes *Transportation Research Record: Journal of the Transportation Research Board* (2136) 2009 pp 67–76

[7] Miller J S Geographical Information Systems Unique Analytic Capabilities for the Traffic Safety Community *Transportation Research Record: Journal of the Transportation Research Board* (1734) 2000 pp 21–28

[8] Pegin P A and Lopashuk V V Usage of GPS Survey Results for Taking the Sun Blinding Effect into Account in Highway Designing *Transport construction* (7) 2008 pp 23–24

[9] Roche J (2000) Geographic Information Systems-Based Crash Data Analysis and the Benefits to Traffic Safety *MTC Transportation Scholars Conference* Ames IA 2000 pp 110–122

[10] Daito Kodama and Ozawa Real Time Accident Risk Information Provision on a Urban Expressway Network: Prediction Model Analysis and Development of a Provision System *Proceedings of the 51st Spring Conference of the Committee of Infrastructure Planning and Management*, 2015

[11] *University of Leeds Driving Simulator* (UoLDS) October 2011http://www.its.leeds.ac.uk/itsresearch/facilities/uolds/

[12] T Kodama et.al. Information service for safe driving by using Road Section Identification Data set (RSIDs) by public-private initiative *Proceedings of the 20th ITS World Congress* Tokyo 2013

[13] Murakami Kurauchi, Yoshii Oonishi, Kawahara Takayama and Hyodo A Study on the Influence of Accident Risk Information to Drivers Choice Behaviors *Proceedings of the 49th Conference of the Committee of Infrastructure Planning and Management* 2014

[14] Kodama T, Yabukami D, Daito T and Ozawa Y 2015 Informing drivers on accident risks: a case of Hanshin Expressway's web-based tool "Safety Drive / Smart Choice"*22nd ITS World Congress* Bordeaux 2015 11 pages

[15] Alexander P, Naveen V, Romesh Hanns de la Fuente-Mella and Luiza M (2015) Traffic and Driving Simulator Based on Architecture of Interactive Motion *The Scientific World Journal* 2015 pp 1–9

[16] Pundt H 2017 Integration of context-information to support spatial decision systems *IADIS International Journal on Computer Science and Information Systems* Vol12 No 2 pp 50-64

[17] Eliseev M, Tomchinskaya T, Lipenkov A and Blinov A Using 3D-modeling Technologies to Increase Road Safety *Transportation Research Procedia* 2017 Vol 20 pp 171-179